Maths

Foundation

Complete Revision and Practice

Rob Kearsley Bullen, Graham Lawlor

Published by BBC Active, an imprint of Educational Publishers LLP, part of the Pearson Education Group Edinburgh Gate, Harlow, Essex CM20 2JE, England

Designed by specialist publishing services ltd

ISBN 978-1-4066-5444-8

Printed in China (CTPSC/03)

First published 2002

This edition 2010

10 9 8 7 6 5 4 3

Minimum recommended system requirements
PC: Windows(r), XP sp2, Pentium 4 1 GHz processor (2 GHz for Vista), 512 MB of RAM (1 GB for Windows Vista), 1 GB of free hard disk space, CD-ROM drive 16x, 16 bit colour monitor set at 1024 x 768 pixels resolution
MAC: Mac OS X 10.3.9 or higher, G4 processor at 1 GHz or faster, 512 MB RAM, 1 GB free space (or 10% of drive capacity, whichever is higher), Microsoft Internet Explorer® 6.1 SP2 or Macintosh Safari™ 1.3, Adobe Flash® Player 9 or higher, Adobe Reader® 7 or higher, Headphones recommended

If you experiencing difficulty in launching the enclosed CD-ROM, or in accessing content, please review the following notes:
1 Ensure your computer meets the minimum requirements. Faster machines will improve performance.
2 If the CD does not automatically open, Windows users should open 'My Computer', double-click on the CD icon, then the file named 'launcher.exe'. Macintosh users should double-click on the CD icon, then 'launcher.osx'
Please note: the eDesktop Revision Planner is provided as-is and cannot be supported.
For other technical support, visit the following address for articles which may help resolve your issues:
http://centraal.uk.knowledgebox.com/kbase/

If you cannot find information which helps you to resolve your particular issue, please email: Digital.Support@pearson.com.
Please include the following information in your mail:
- Your name and daytime telephone number.
- ISBN of the product (found on the packaging.)
- Details of the problem you are experiencir
- Details of your computer (operating systen

Contents

Shape, space and measures

Handling data

* Only available in the CD-ROM version of the book

Exam board specification map

Provides a quick and easy overview of the topics you need to study for the examinations you will be taking. This map is based on the Mathematics National Curriculum for England only.

Topics	Page	AQA A linear	AQA B modular	Edexcel linear	Edexcel modular	OCR A linear	OCR B modular
Number							
The decimal number system	2	✓	✓	✓	✓	✓	✓
Order of operations	4	✓	✓	✓	✓	✓	✓
Negative numbers	6	✓	✓	✓	✓	✓	✓
Factors and multiples	8	✓	✓	✓	✓	✓	✓
Working with fractions	10	✓	✓	✓	✓	✓	✓
Fractions, decimals and percentages	12	✓	✓	✓	✓	✓	✓
Powers and roots	14	✓	✓	✓	✓	✓	✓
Standard index form	16	✓	✓	✓	✓	✓	✓
Ration and proportion	18	✓	✓	✓	✓	✓	✓
Percentage calculations	20	✓	✓	✓	✓	✓	✓
Algebra							
Algebraic expressions	22	✓	✓	✓	✓	✓	✓
Formulae and substitution	24	✓	✓	✓	✓	✓	✓
Solving equations	26	✓	✓	✓	✓	✓	✓
Trial and improvement	28	✓	✓	✓	✓	✓	✓
Rearranging formulae	30	✓	✓	✓	✓	✓	✓
Using brackets in algebra	32	✓	✓	✓	✓	✓	✓
Multiplying bracketed expressions	34	✓	✓	✓	✓	✓	✓
Inequalities	36	✓	✓	✓	✓	✓	✓
Number patterns and sequences	38	✓	✓	✓	✓	✓	✓
Sequences and formulae	40	✓	✓	✓	✓	✓	✓
Co-ordinates	42	✓	✓	✓	✓	✓	✓
Lines and equations	44	✓	✓	✓	✓	✓	✓
Quadratic graphs	46	✓	✓	✓	✓	✓	✓

Introduction

How to use GCSE Bitesize Complete Revision and Practice

Begin with the CD-ROM. There are five easy steps to using the CD-ROM – and to creating your own personal revision programme. Follow these steps and you'll be fully prepared for the exam without wasting time on areas you already know.

Topic checker

Step 1: Check

The Topic checker will help you figure out what you know – and what you need to revise.

Revision planner

Step 2: Plan

When you know which topics you need to revise, enter them into the handy Revision planner. You'll get a daily reminder to make sure you're on track.

Revise

Step 3: Revise

From the Topic checker, you can go straight to the topic pages that contain all the facts you need to know.

- Give yourself the edge with the Web*Bite* buttons. These link directly to the relevant section on the BBC Bitesize Revision website.

- Audio*Bite* buttons let you listen to more about the topic to boost your knowledge even further. *

Step 4: Practise

Check your understanding by answering the Practice questions. Click on each question to see the correct answer.

Step 5: Exam

Are you ready for the exam? Exam*Bite* buttons take you to an exam question on the topics you've just revised. *

*** Not all subjects contain these features, depending on their exam requirements.**

Interactive book

You can choose to go through every topic from beginning to end by clicking on the Interactive book and selecting topics on the Contents page.

Exam questions

Find all of the Exam questions in one place by clicking on the Exam questions tab.

Last-minute learner

The Last-minute learner gives you the most important facts in a few pages for that final revision session.

You can access the information on these pages at any time from the link on the Topic checker or by clicking on the Help button. You can also do the Tutorial which provides step-by-step instructions on how to use the CD-ROM and gives you an overview of all the features available. You can find the Tutorial on the Home page when you click on the Home button.

Other features include:

Click on the draw tool to annotate pages. N.B. Annotations cannot be saved.

Click on Page turn to stop the pages turning over like a book.

Click on the Single page icon to see a single page.

Click on this arrow to go back to the previous screen.

Contents

Click on Contents while in the Interactive book to see a contents list in a pop-up window.

Click on these arrows to go backward or forward one page at a time.

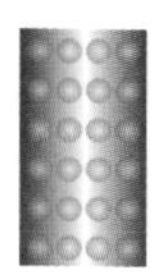

Click on this bar to switch the buttons to the opposite side of the screen.

Click on any section of the text on a topic page to zoom in for a closer look.

N.B. You may come across some exercises that you can't do on-screen, such as circling or underlining, in these cases you should use the printed book.

About this book

Use this book whenever you prefer to work away from your computer. Before you begin, use the Topic checker that follows this Introduction to help you figure out what you know – and what you need to revise. The book consists of two main parts:

A set of double-page spreads, covering the essential topics for revision from each of the curriculum areas. Each topic is organised in the following way:

- A summary of the main points and an introduction to the topic.
- Lettered section boxes cover the important areas within each topic.
- Key facts are clearly highlighted – these indicate the essential information in a section or give you tips on answering exam questions.
- Practice questions at the end of each topic – a range of questions to check your understanding.

A number of special sections to help you consolidate your revision and get a feel for how exam questions are structured and marked. These extra sections towards the back of this book will help you check your progress and be confident that you know your stuff. They include:

- A selection of exam-style questions and worked model answers and comments to help you get full marks.
- Complete the facts – check that you have the most important ideas at your fingertips.
- Last-minute learner – the most important facts in just a few pages.

About your exam

Get organised

You need to know when your exams are before you make your revision plan. Check the dates, times and locations of your exams with your teacher, tutor or school office.

On the day

Aim to arrive in plenty of time, with everything you need: several pens, pencils, a ruler, and possibly mathematical instruments, a calculator, or a language dictionary, depending on the exam subject.

On your way or while you're waiting, read through your Last-minute learner.

In the exam room

When you are issued with your exam paper, you must not open it immediately. However, there are some details on the front cover that you can fill in (your name, centre number, etc.) before you start the exam itself. If you're not sure where to write these details, ask one of the invigilators (teachers supervising the exam).

When it's time to begin writing, read each question carefully. Remember to keep an eye on the time.

Finally, don't panic! If you have followed your teacher's advice and the suggestions in this book, you will be well-prepared for any question in your exam.

Techniques to help you remember

There are things that you can do to enhance your memory as you revise.

- **Shapes** – Suppose that you had to remember five different types of numbers; for example, primes, squares, cubes, triangular numbers, and the Fibonacci sequence.

  One easy way to remember these five facts is to draw them around a pentagon:

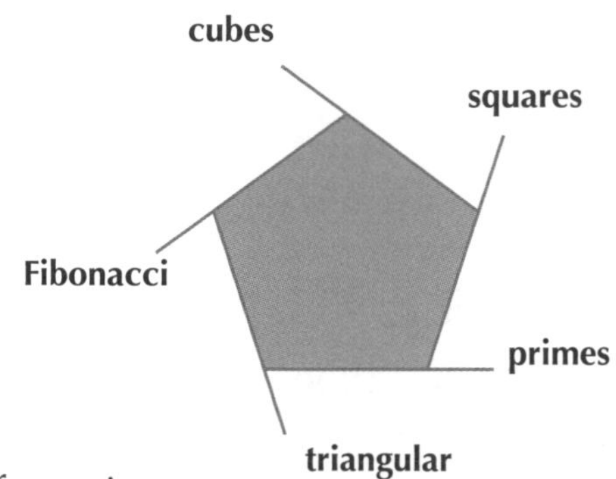

  You can then add extra diagrams or written facts to the pentagon. The act of drawing the pentagon and labelling it causes you to organise your thoughts in the right way for the topic you're working on; for example, the fact that the pentagon has five sides means that you know you have to remember five facts.

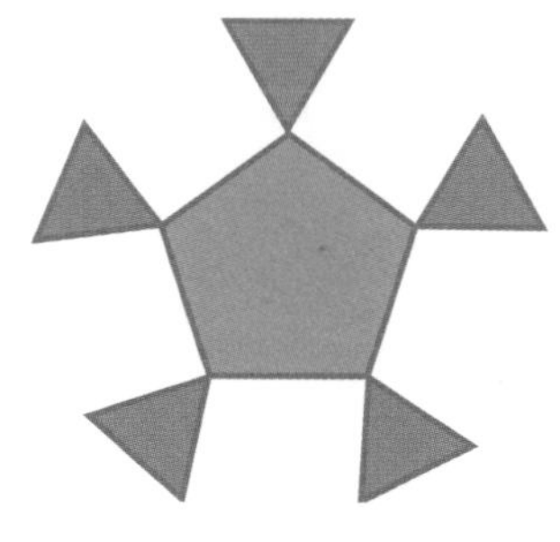

  Suppose you had five categories of information to memorise and each category had three sets of facts. You could develop your shape memory-jogger to look like the one we've drawn on the left.

  Here, the five facts can be listed around the pentagon and each subsection of three facts can be listed around each small triangle. If some of the sub-sections have only one fact, instead of using a triangle, use a circle. The shape memory-jogger is fun, creative and uses both sides of your brain.

- **Chunking** – This means grouping facts together. For instance, one of the easiest ways to remember phone numbers is to group the digits in threes. So, if you can group material, it will make it easier to remember.

- **Acronyms** – Acronyms are 'words' made from the initial letters of facts you need to remember. You can find a number of these (such as BIDMAS) in the book already, but you can easily make your own.

- **Phrases** – Sometimes you can remember a fact using a humorous phrase. For example, people remember that < means 'less than' and > means 'greater than' using the phrase 'the crocodile's mouth eats the biggest thing it can'.

- **Images** – Making images is creative, involving you more actively in the process of revising, helping you to remember things easily. In your revision notes, make pictures and images as often as you can.
  Good images:
  – are colourful
  – are lively and dynamic
  – can make you laugh
  – often have exaggerated aspects to them.

  Here's an example illustrating 'along the corridor before you climb the stairs' for co-ordinates:

Topic checker

Go through these questions after you've revised a group of topics, putting a tick if you know the answer.

You can check your answers on pages xv–xix.

>> Number

1 Work out the equivalent fractions:

(a) $\frac{1}{6} = \frac{}{18}$ (b) $\frac{3}{4} = \frac{15}{}$

2 Change these improper fractions to mixed numbers:

(a) $\frac{21}{4}$ (b) $\frac{83}{9}$ (c) $\frac{61}{10}$

3 Change these mixed numbers to improper fractions:

(a) $5\frac{1}{3}$ (b) $22\frac{3}{4}$ (c) $13\frac{1}{8}$

4 What is the denominator of a fraction?

5 What is the numerator of a fraction?

6 Write 60 000 in standard form.

7 What is 0.000 000 000 075 in standard form?

8 Change 4.6×10^3 into ordinary form.

9 Use your calculator to find $31 \div 13$. Round the answer to:
(a) 1 decimal place (b) 3 decimal places (c) 3 significant figures.

10 Find $4 \times 10^8 + 5 \times 10^7$ without using a calculator.

11 What is a prime factor?

12 Express 100 as a product of its prime factors.

13 What is the next cube number after 27?

14 Write down the first ten terms of the Fibonacci sequence.

15 What are the factors of 100?

16 What is a prime number?

17 Express $\frac{t^6}{t^2}$ as a single power of t.

18 Evaluate 3^5.

19 Evaluate 15^0.

20 Express $(y^5)^2$ as a single power of y.

21 A school roll of 800 students increased by 15% between September and Christmas. It then increased further by another 5%. How many students are now on the roll?

22 For the numbers 48 and 40, find:

(a) the highest common factor (b) the lowest common multiple.

23 Write the ratio 6 : 10 in: (a) simplest form (b) the form $n : 1$ (c) the form $1 : n$.

24 Divide £150 in the ratio 3 : 2 : 1.

25 Calculate the VAT on an item costing £30 (calculate VAT at 17.5%).

26 Add brackets to this calculation to make it correct:

$10 - 3 \times 2 + 5 + 1 = 50$

>> Algebra

27 Solve $3x + 5 = 26$.

28 Solve $10 - \frac{1}{4}k = 9$.

29 Solve $5 - 8p = 5p - 21$.

30 Solve $5 - 8y = 37$.

31 Make x the subject of $\frac{p}{x} = m$.

32 Make x the subject of $6 - fx = 4x + j$.

33 Make x the subject of $\frac{ax}{b} = c$.

34 Make c the subject of $E = mc^2$.

35 Solve the inequality $5x - 4 > 6$.

36 Solve the inequality $-9f < 81$.

37 What is special about the co-ordinates of points in the first quadrant?

38 In which quadrant is the x co-ordinate negative and the y co-ordinate positive?

Topic checker

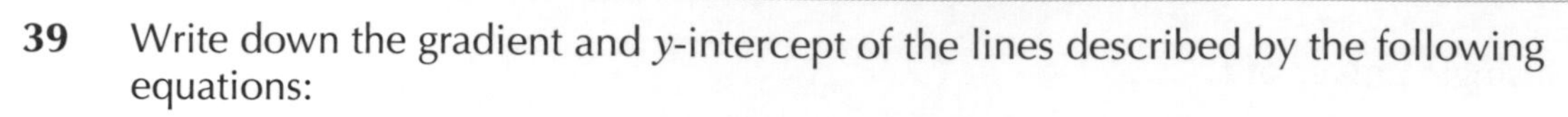

39 Write down the gradient and y-intercept of the lines described by the following equations:

(a) $y = 9x + 3$ (b) $y = 4 - 3x$ (c) $y = 6x + 5$ (d) $y = 5 - 7x$.

40 Rearrange these equations to make y the subject of the equation and then write down the gradient and the y-intercept:

(a) $2x + 3y = 4$ (b) $5x - 7y = 12$.

41 What can you say about the gradient of this line?

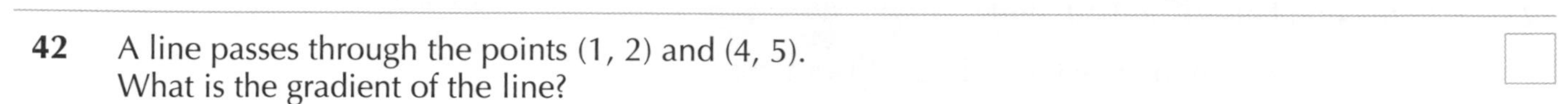

42 A line passes through the points (1, 2) and (4, 5). What is the gradient of the line?

43 Find the equation of the line that passes through the point (0, 5) that is parallel to the line $y = 2x + 1$.

44 Solve $3(x + 5) = 42$.

45 Factorise: (a) $100v^2 + 20vc$ (b) $169g^2 - 225h^2$.

46 Expand and simplify: (a) $(x + 3)(x + 5)$ (b) $4x(x + 5)(x + 6)$.

47 What is an identity?

48 Find the two missing numbers from this sequence: 2, 4, 10, ?, 34, ?, …

49 Find the two missing numbers from this sequence: 10, 5, ?, ?, 0.625, …

50 Find the first four terms of these sequences:

(a) $u_n = 3n + 2$ (b) $u_n = 5^n$.

51 Find formulae for u_n to describe each of these sequences:

(a) 2, 5, 8, 11 (b) 7, 12, 17, 22.

52 $S = \frac{1}{2}at^2$

Find t, when $s = 90$ and $a = 5$.

53 $m = \sqrt{\frac{d}{2\pi}}$

Find d, when $m = 100$.

54 $A = \pi r^2$

When $A = 169$, find r.

55 Draw the graph of $y = x^2 - 5x + 4$.

56 Draw the graph of $y = x^2 + 1$ and use the graph to solve $x^2 + 1 = 3$.

57	By trial and improvement, find a solution to $z^3 - 10z = 1$, correct to 1 dp.	☐
58	If a measurement is given as 30 ml, correct to the nearest ml, what range of values is possible?	☐
59	A film lasts 1 hour 48 minutes and finishes at 23:15. What time did it start?	☐
60	If a journey of 75 km takes 90 minutes, what is the average speed?	☐

>> Shapes, space and measures

61 Find the area of this triangle. ☐

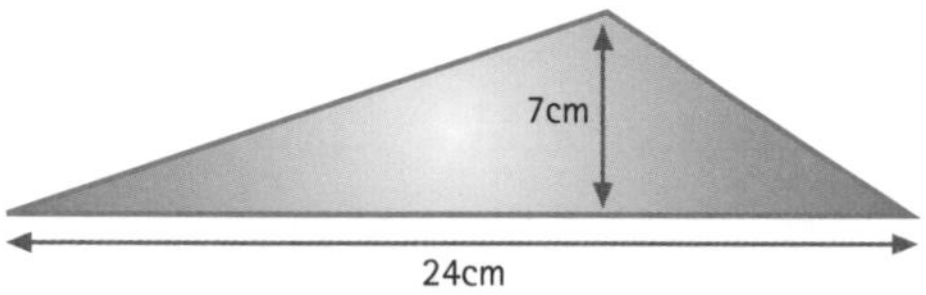

7cm
24cm

62 Find the volume of a triangular prism of length 10 cm, and triangle base length of 3 cm. The prism is 4 cm in height. ☐

63 Find the volume of a cylinder of base radius 15 cm and height 14 cm. ☐

64 What is a sector of a circle? ☐

65 Find the area of a circle with a diameter of 25 cm. ☐

66 A circle has an area of 100 cm². Find its radius. ☐

67 What is the circumference of a circle with a diameter of 50 cm? ☐

68 A circle has a circumference of 24 cm. What is its radius? ☐

69 Which quadrilaterals have two lines of symmetry? ☐

70 Find the angles in this triangle. ☐

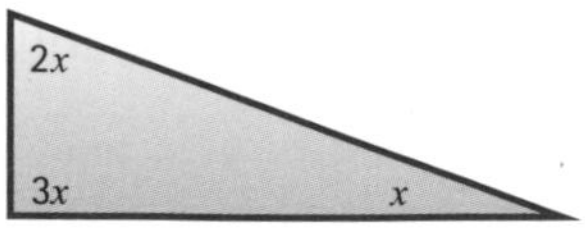

$2x$
$3x$
x

71 What are the three angle relationships created by parallel lines, and which of these link equal angles? ☐

72 What is Pythagoras' Rule? ☐

73 Calculate x. ☐

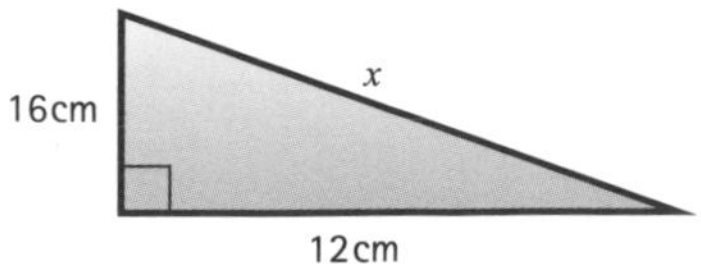

16cm
x
12cm

74 Find the length of m. ☐

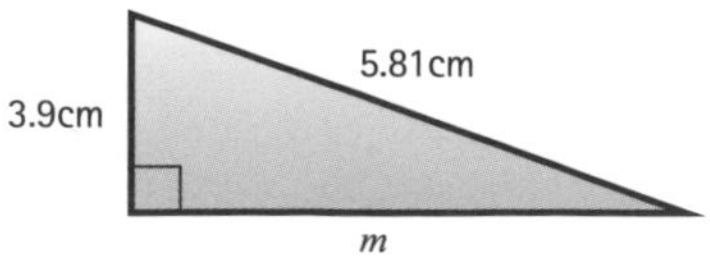

3.9cm
5.81cm
m

Topic checker

75 A square has diagonals of 30 cm. Find the length of one of the sides.

76 Find the length of the diagonal of a rectangle that measures 10 cm by 40 cm.

77 What locus is equidistant from two points?

78 Draw a net for a cuboid 5 cm by 2 cm by 1 cm.

79 What information do you need to give to describe:

(a) a translation (b) a reflection (c) a rotation (d) an enlargement?

>> Handling data

80 The following data for the number of cars in a car park was collected:

Type of car	Ford	Volvo	Skoda	Toyota
Number of cars	55	25	15	80

Draw a clearly labelled pie chart to show this information.

81 What is correlation?

82 If the points are scattered closely around the line of best fit on a scatter diagram, what does that say about the correlation?

83 If the points are scattered very loosely so that there is no clear line of best fit on a scatter diagram, what does that say about the correlation?

84 If the points are scattered from top left to bottom right on a scatter diagram, what does that say about the correlation?

85 Define these terms exactly:

(a) mean (b) median (c) mode (d) range.

86 What two statistical values should you use to compare two sets of data?

87 The probability of an outcome happening and the probability of it not happening add up to what?

88 The red cards in a pack are removed. One card is picked at random. What is the probability of it being:

(a) the Ace of Spades (b) a Club (c) not a Diamond (d) the King of Hearts?

Topic checker answers

>> Number

1	**(a)** $\frac{1}{6} = \frac{3}{18}$ **(b)** $\frac{3}{4} = \frac{15}{20}$
2	**(a)** $\frac{21}{4} = 5\frac{1}{4}$ **(b)** $\frac{83}{9} = 9\frac{2}{9}$ **(c)** $\frac{61}{10} = 6\frac{1}{10}$
3	**(a)** $5\frac{1}{3} = \frac{16}{3}$ **(b)** $22\frac{3}{4} = \frac{91}{4}$ **(c)** $13\frac{1}{8} = \frac{105}{8}$
4	The lower part of the fraction.
5	The upper part of the fraction.
6	6×10^4
7	7.5×10^{-11}
8	1.72
9	**(a)** 2.4 **(b)** 2.385 **(c)** 2.38
10	4.5×10^8
11	A factor of a number that is itself a prime.
12	$2^2 \times 5^2$
13	64
14	1, 1, 2, 3, 5, 8, 13, 21, 34, 55
15	1, 2, 4, 5, 10, 20, 25, 50, 100
16	A number that has only two factors and the factors are different. The factors are 1 and the number itself.
17	t^4
18	243
19	1
20	y^{10}
21	966 students
22	**(a)** 8 **(b)** 240
23	**(a)** 3 : 5 **(b)** 0.6 : 1 **(c)** 1 : $1\frac{2}{3}$

Topic checker answers

24 £75 : £50 : £25

25 £5.25

26 $(10 - 3) \times (2 + 5) + 1 = 50$

>> Algebra

27 $x = 7$

28 $k = 4$

29 $p = 2$

30 $y = -4$

31 $x = \frac{p}{m}$

32 $x = \frac{6-j}{f+4}$

33 $x = \frac{cb}{a}$

34 $c = \sqrt{\frac{E}{m}}$

35 $x > 2$

36 $f > -9$

37 They are both positive.

38 The second quadrant.

39 **(a)** gradient = 9, y-intercept = 3
(b) gradient = −3, y-intercept = 4
(c) gradient = 6, y-intercept = 5
(d) gradient = −7, y-intercept = 5

40 **(a)** $y = \frac{4-2x}{3}$ gradient $= \frac{-2}{3}$ y-intercept $= \frac{4}{3}$

(b) $y = \frac{5x-12}{7}$ gradient $= \frac{5}{7}$ y-intercept $= 1\frac{5}{7}$

41 It is negative.

42 1

43 $y = 2x + 5$

44 $x = 9$

45 **(a)** $20v(5v + c)$
(b) $(13g + 15h)(13g - 15h)$

46 (a) $x^2 + 8x + 15$
(b) $4x^3 + 44x^2 + 120x$

47 A statement that is true for all values of the variables involved in it, e.g. $2x = x + x$

48 20 and 52

49 2.5 and 1.25

50 (a) 5, 8, 11, 14 (b) 5, 25, 125, 625

51 (a) $u_n = 3n - 1$ (b) $u_n = 5n + 2$

52 $t = 6$

53 62 831.86

54 7.33

55

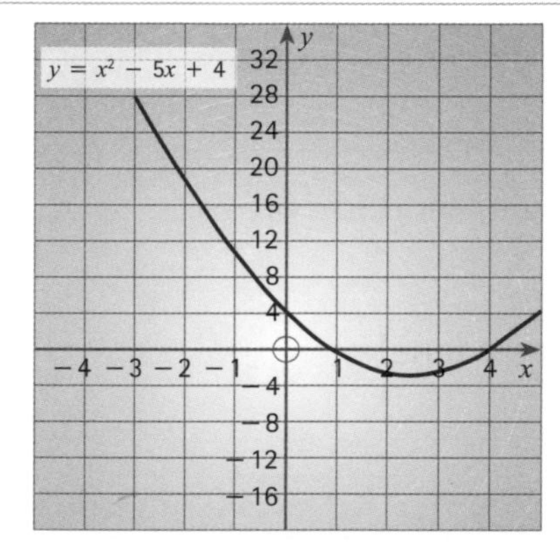

56

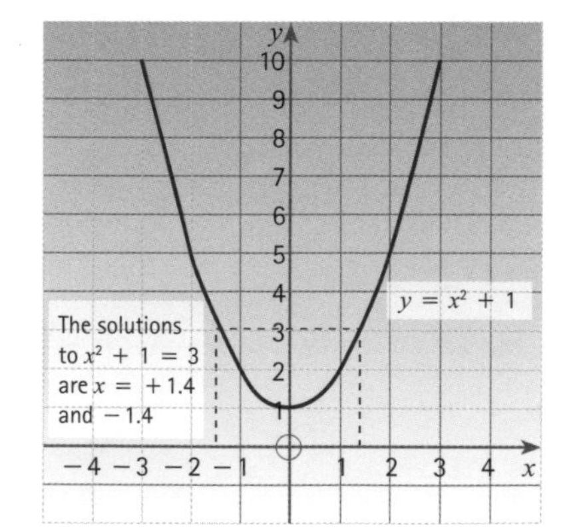

57 $z = -0.1$, or 3.2

58 29.5 ml < measurement ⩽ 30.5 ml

59 21:27

60 50 km/h

Topic checker answers

>> Shape, space and measures

61 84 cm^2

62 60 cm^3

63 9896 cm^3

64 A slice, like a slice of a circular pie.

65 490.87 cm^2

66 5.64 cm

67 157.1 cm

68 3.82 cm

69 Rectangle, rhombus

70 $x = 30°$, $2x = 60°$, $3x = 90°$

71 Corresponding angles (equal), alternate angles (equal), allied angles (add to 180°)

72 $h^2 = a^2 + b^2$
More formally it is 'the square of the hypotenuse of a right-angled triangle is equal to the sum of the squares of the other two sides'.

73 $x = 20$ cm

74 $m = 4.31$ cm to 2 dp

75 21.21 cm to 2 dp

76 41.23 cm

77 The perpendicular bisector of the points.

78

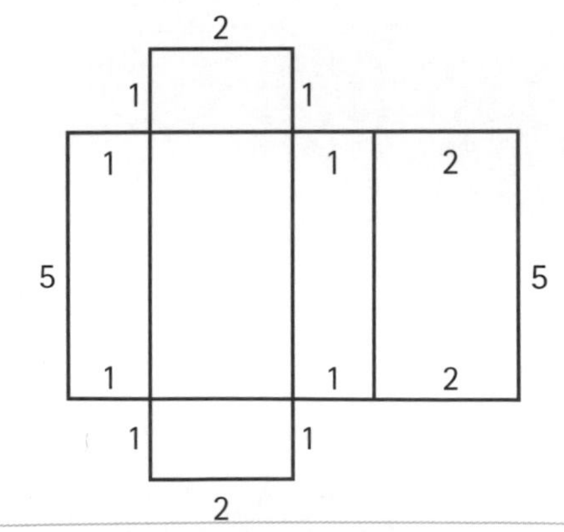

79 **(a)** a vector
(b) the position of the mirror line
(c) the angle/direction and the centre of rotation
(d) the scale factor and the centre of enlargement

>> Handling data

80 The pie chart should be divided up:
Ford (113°), Volvo (51°),
Skoda (31°), Toyota (165°).

81 Correlation is a relationship between two sets of data. The classic example is smoking. There is a close correlation between the numbers of people who smoke and those who later develop lung cancer.

82 There is a strong correlation.

83 There is no correlation.

84 It is a negative correlation.

85 **(a)** The mean is the arithmetical mean. To find the mean, add up all of the data and divide this answer by the number of data items.
(b) The median is the middle number when the data is placed in order of size.
(c) The mode is the item of data that occurs the most often.
(d) The range is the difference between the highest and the lowest values in a data set.

86 An average (mean, median or mode) and the range.

87 1

88 **(a)** $\frac{1}{26}$ **(b)** $\frac{1}{2}$ **(c)** 1 **(d)** 0

The decimal number system

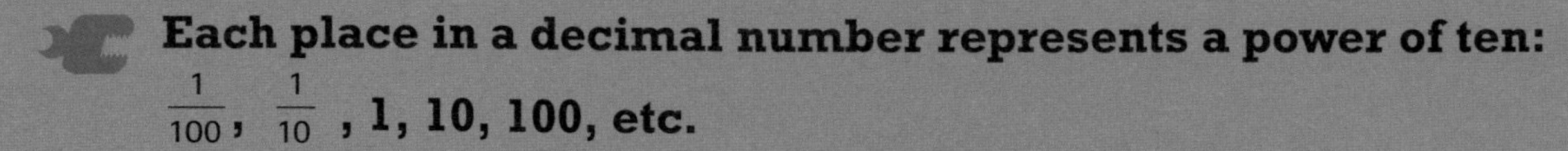

Each place in a decimal number represents a power of ten: $\frac{1}{100}$, $\frac{1}{10}$, 1, 10, 100, etc.

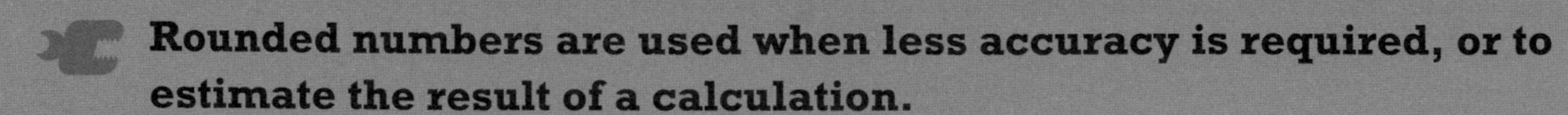

Rounded numbers are used when less accuracy is required, or to estimate the result of a calculation.

A Place value

1. The system of numbers we use is called the **decimal system** because it is based on the number ten. Numbers in the decimal system are composed of **digits**. The position of a digit in a number controls its **place value**.

2. A decimal point separates the whole and fractional parts of a number.

>> **key fact** **Digits to the left of the decimal point represent whole numbers; those to the right, fractions.**

3. The digit **2** means something different in each of these numbers:

Number	Position of '2'	Value of '2'
1**2**5	tens	20
2130559	millions	2000000
3.**2**8	tenths	$\frac{2}{10}$
0.00**2**6	thousandths	$\frac{2}{1000}$

B Decimal calculations

1. **key fact** **To multiply a number by 10, 100 or 1000, simply move its digits 1, 2 or 3 places to the left:**

$$45.67 \times 10 = 456.7$$
$$45.67 \times 100 = 4567$$
$$45.67 \times 1000 = 45\,670.$$

>> **key fact** **To divide by 10, 100 or 1000, move digits 1, 2 or 3 places to the right:**

$$45.67 \div 10 = 4.567$$
$$45.67 \div 100 = 0.4567$$
$$45.67 \div 1000 = 0.045\,67.$$

Multiplying by 0.1 is the same as dividing by 10.

2. You can use the result of a whole number multiplication to find the answer to many decimal multiplications:

$12 \times 2 = 24$,
so $12 \times 0.2 = 2.4$
$12 \times 0.02 = 0.24$
and $1.2 \times 2 = 2.4$
$1.2 \times 0.2 = 0.24$
$1.2 \times 0.02 = 0.024$, etc.

If one of the numbers is made ten times smaller, the answer will be ten times smaller.

3. A similar rule works with division, but if the number you are dividing by gets smaller, the answer gets bigger:

$12 \div 2 = 6$,
so $12 \div 0.2 = 60$
$12 \div 0.02 = 600$
and $1.2 \div 2 = 0.6$
$1.2 \div 0.2 = 6$
$1.2 \div 0.02 = 60$, etc.

C Rounded numbers

1 Sometimes an answer to a question will contain a lot of decimal digits. Try 10 ÷ 7 on your calculator – it should fill the display completely. In practice, you don't always need an answer this accurate, so you may need to **round** it. This changes the value slightly, but makes it much easier to read.

2 *Rounding to the nearest 10, 100, etc.*

The number 24 is closer to 20 than it is to 30, so 24 rounded to the nearest ten is 20. 25 is exactly halfway. In a 'halfway' situation, you always round **up**. So 25 rounded to the nearest ten is 30.

>> **key fact** **If the 'next' digit is 5 or over, round up.**

3 *Rounding to a given number of decimal places (dp)*

Suppose you needed to round 1.428 571 4 to 2 dp.

- Split the number after the required number of decimal places: 1.42 | 85714
- Look at the first digit after the split. It's 8. This is over 5, so you round the number **up** to 1.43.

4 *Rounding to a given number of significant figures (sf)*

Here, the type of rounding depends on the size of the number. Count digits from the first non-zero digit in the number, and split after this. The table shows some different cases.

Number	Round to	Split number	Type of rounding	Rounded number
6415	2 sf	64 | 15	nearest hundred	6400
0.066 666 66	3 sf	0.0666 | 6666	4 dp	0.0667
84.9	1 sf	8 | 4.9	nearest ten	80

D Estimating answers

1 Sometimes it is useful to check a calculation by making an **estimate** of the answer.

>> **key fact** **Round the numbers in the calculation to 1 sf first.**

2 To estimate the answer to $\frac{235 \times 7.81}{38.33}$, round the numbers so the calculation becomes $\frac{200 \times 8}{40}$.

Then calculate with these numbers: $\frac{200 \times 8}{40} = \frac{1600}{40} = 40$.

3 The actual answer using the original numbers is $\frac{1835}{38.33} = 47.88$ to 2 dp. This is close to the estimate, so you can be confident it's correct.

>> practice questions

1 Do a whole number calculation first, then use the result to answer the question.

(a) 2.5×3 **(b)** 0.3×1.2 **(c)** $6.4 \div 8$ **(d)** $1.44 \div 0.03$

2 Round each number in the three different ways given.

(a) 61.25; (i) nearest unit (ii) 1 dp (iii) 2 dp

(b) 588.621; (i) 1 sf (ii) 2 sf (iii) 4 sf

3 Estimate the answer to each calculation, then find the exact answer, rounded to 3 sf.

(a) $(31.42 - 15.7) \times 2.25$ **(b)** $\frac{13.7 \times 35.1}{14}$ **(c)** 5.89^2 **(d)** $\frac{3.57}{1.81} \times \frac{2.26}{4.009}$

Order of operations

There is a standard order in which you carry out operations. Brackets can be used to change this order.

Use BIDMAS to remember the order.

A 'Stronger' and 'weaker' operations

1. Mathematical **operations** allow numbers to be combined in a calculation.

   The usual operations are:

   - addition and subtraction
   - multiplication and division
   - powers and roots.

2. When you see more than one operation in a calculation, there is a standard order you must use to work out the answer.

>> **key fact** **Multiplication and division should be done before addition and subtraction as they are 'stronger' operations.**

$2 + 4 \times 5$ means $2 + 20 = 22$, **not** $6 \times 5 = 30$.

>> **key fact** **Powers and roots are the 'strongest' and should always be done first.**

3×2^2 means $3 \times 4 = 12$, not $6^2 = 36$.

3. When operations of the same 'strength' are mixed, you simply work through them in order.

   So $10 - 3 + 2 - 5 = 4$.

B Brackets

1. The only way to change the order in a calculation is to use **brackets**. The part of the calculation that occurs in the bracket **must** be evaluated first.

   $(2 + 4) \times 5$ means $6 \times 5 = 30$.

   $(3 \times 2)^2$ means $6^2 = 36$.

   Sometimes, you may see $8 \times (9 - 3)$ written as $8(9 - 3)$, as in algebra.

2. Some calculations contain 'hidden' brackets.

   In this one, fraction-style notation has been used to show division.

   To work out $\frac{23.85 + 46.2}{30}$, you actually calculate $(23.85 + 46.2) \div 30$.

   A square root sign can also 'hide' a bracket.

   $\sqrt{3^2 + 4^2}$ means the square root of $(3^2 + 4^2)$, **not** $\sqrt{3^2} + 4^2$.

3. Occasionally, you may find brackets inside brackets. This is called **nesting**. Always start with the innermost bracket in the calculation.

   $((23 + 37) \div 4)^2 = (60 \div 4)^2 = 15^2 = 225$

C Calculators

1 A scientific calculator obeys the correct rules for operations. The display may show answers to some of the steps in a calculation as you work. Don't be put off by this! A standard calculator will probably do calculations in the order they are entered, and will give incorrect results for some calculations.

2 Your calculator should have a pair of bracket keys [( )] . These alter the order of operations in a calculation in exactly the way you would expect.

3 You can also use your calculator's **memory** to store part of a long calculation. Make sure you know how these functions work – read your calculator's instruction manual!

The memory keys should look like these:

Min M+  

Or these:  RCL CLR

D BIDMAS

1 **key fact** **You can use the made-up word BIDMAS to help you remember the order of operations. Each letter stands for an operation.**

B brackets

I indices (this covers powers and roots)

D division }

M multiplication }

A addition }

S subtraction }

practice questions

1 Use the correct order of operations to calculate these.

(a) $10 + 3 \times 5$ **(b)** $12 - 2^2$ **(c)** $24 \div 3 + 7$

(d) 5×2^2 **(e)** $4 \times 10 + 10 \times 9$ **(f)** $3.5 + 15 \div 2$

2 Carry out these calculations involving brackets.

(a) $(6 + 2) \times 4$ **(b)** $10 \div (23 - 21)$ **(c)** $2 \times (2 + 2) \times 2$

(d) $10(3.6 + 2.1)$ **(e)** $(3 \times (2 + 1))^2$ **(f)** $(12 \div 3)^3$

3 Some of these calculations need brackets to make them correct. Add brackets where required.

(a) $4 + 2 \times 3 = 18$ **(b)** $12 \div 6 - 3 = 4$ **(c)** $5 + 6 \times 6 + 5 = 46$

(d) $5 + 6 \times 6 + 5 = 121$ **(e)** $4 + 3^2 = 13$ **(f)** $1 + 2 \times 3 \div 4 + 5 = 1$

4 Evaluate these using your calculator. Round your answer to 3 significant figures if needed.

(a) $(2.15 + 4.22) \times 55$ **(b)** $\frac{1000}{16.3 \times 41.45}$ **(c)** $\sqrt{4^2 - 2^2}$

(d) $(5^2 - 40)^2$ **(e)** $(14.1 - (6.25 + 8.15)^2)^2$ **(f)** $(-10)^3$

Negative numbers

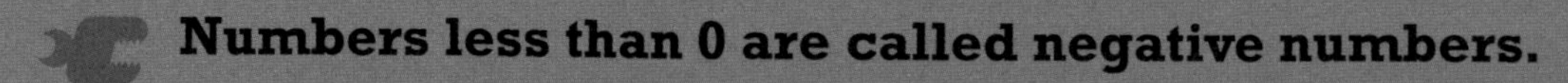

- **Numbers less than 0 are called negative numbers.**

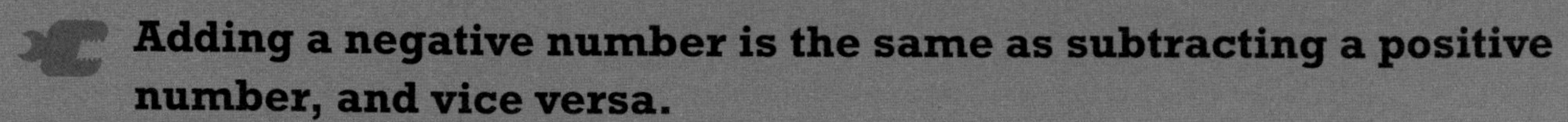

- **Adding a negative number is the same as subtracting a positive number, and vice versa.**
- **Multiplying or dividing numbers with opposite signs gives a negative answer.**

A The number line

1. Positive numbers are greater than zero (> 0). Negative numbers are less than zero (< 0).
2. Positive and negative numbers can be represented easily on a number line.

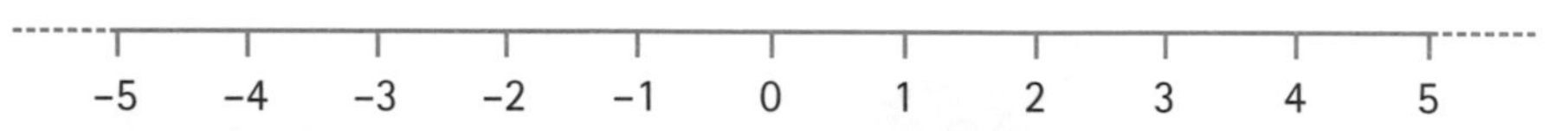

B Addition and subtraction

1. **key fact** **Adding a negative number gives the same result as subtracting a positive number.**

Examples: $4 + (-3) = 4 - 3 = 1$ *brackets are used to make it clearer*

$3 + (-7) = 3 - 7 = -4$ *you could use a number line to help*

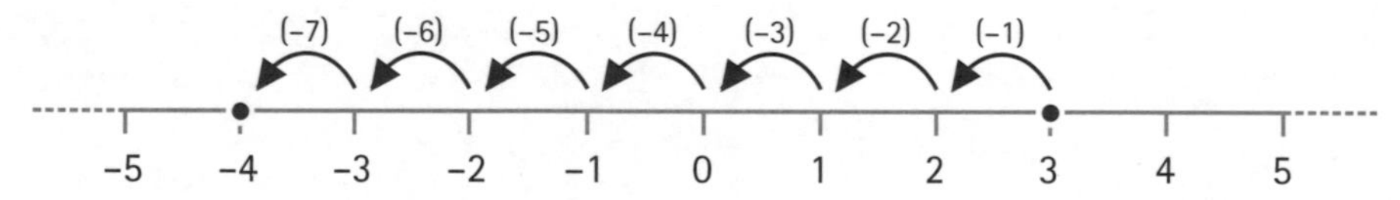

Another way to do the last example is to notice that when you swap the numbers in a subtraction, you change the sign of the answer: $7 - 3 = 4$ and so $3 - 7 = -4$

This can be useful when larger numbers are involved: $25 + (-42) = 25 - 42$

$42 - 25 = 17$

so $25 - 42 = -17$

2. **key fact** **Subtracting a negative number gives the same result as adding a positive number.**

Examples:

$2 - (-4) = 2 + 4 = 6$

$(-12) - (-36) = (-12) + 36 = 24$ *use a 'rough' number line to help*

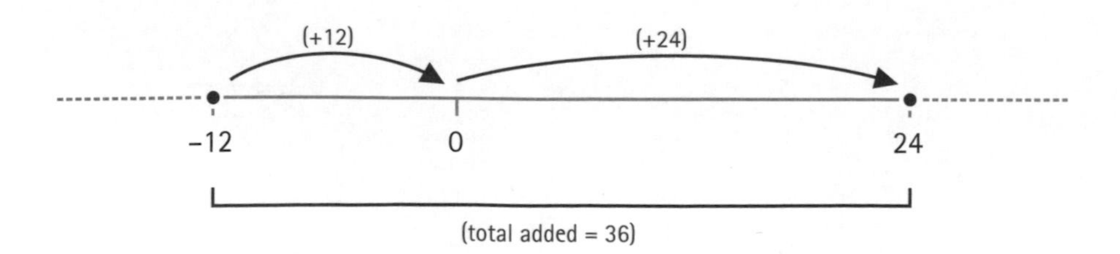

C Multiplication and division

1 The rules for multiplication and division are very simple:

>> **key fact** **negative × positive = negative: negative × negative = positive**
negative ÷ positive = negative: negative ÷ negative = positive

Examples:

$3 \times (-5) = -(3 \times 5) = -15$ $(-4) \times (-10) = 4 \times 10 = 40$

$(-16) \div 2 = -(16 \div 2) = -8$ $(-39) \div (-3) = 39 \div 3 = 13$

2 You can remember this with the 'word' **SPON**:

'**S**ame (signs) **P**ositive, **O**pposite (signs) **N**egative'. Otherwise, use this table:

× / ÷	+	−
+	+	−
−	−	+

D Squaring and cubing

1 **key fact** **The square of a negative number is positive.**

Example: $(-7)^2 = (-7) \times (-7) = 49$

2  **The cube of a negative number is negative.**

Example: $(-2)^3 = (-2) \times (-2) \times (-2)$
$= 4 \times (-2) = -8$

E Calculators

1 Your calculator has a key like this [(−)] for entering negative numbers. It will carry out calculations using negative numbers and give the correct results in every case *except* the one described in the next section.

2 There is a programming error in some calculators that can cause problems with operations using negative numbers.
Try [(−)] [7] [x^2] to calculate the square of −7.
Now try [[(] [(−)] [7] [)]] [x^2].
The correct answer is 49 (see section D) — does your calculator get it right?

>> practice questions

1 (a) At 4 pm on Monday, the temperature was 6° C. By midnight, it had fallen to −3° C. By how many degrees did it fall?

(b) By 1 pm on Tuesday, the temperature had risen by 12° C. What was it then?

2 Work these out without a calculator.

(a) $3 - 10$ **(b)** $4 + (-2)$ **(c)** $(-5) + 9$ **(d)** $(-5) + (-9)$
(e) $4 \times (-8)$ **(f)** $(-3) \times (-7)$ **(g)** $(-24) \div (-3)$ **(h)** $(-10)^2$

3 Use your calculator to find these.

(a) $138 - 272$ **(b)** $67 + (-125)$ **(c)** $(-3320) + 2671$ **(d)** $(-6.93) - (-4.63)$

(e) $(-255) \times 30$ **(f)** $27 \div (-0.54)$ **(g)** $(-8) \div (-200)$ **(h)** $(-0.05)^2$

Factors and multiples

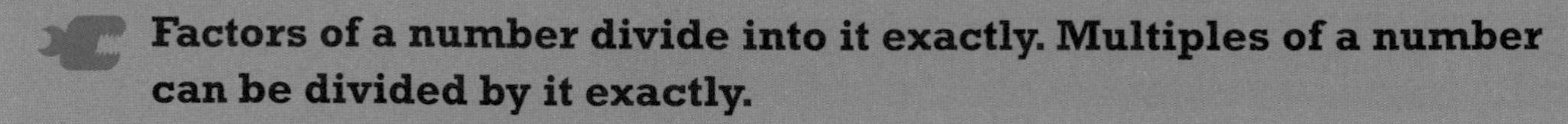

Factors of a number divide into it exactly. Multiples of a number can be divided by it exactly.

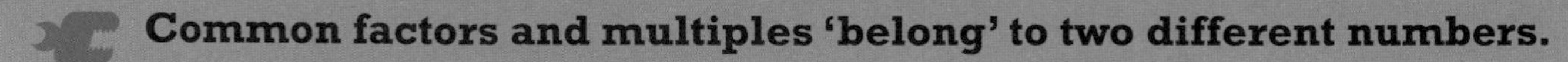

Common factors and multiples 'belong' to two different numbers.

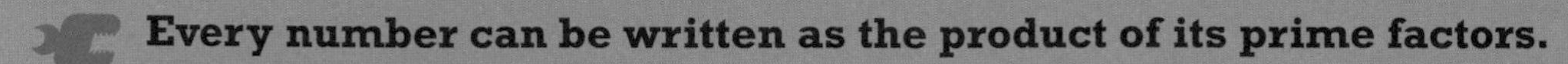

Every number can be written as the product of its prime factors.

A Factors

1 **key fact** **The factors of a whole number are the numbers that divide into it exactly.**

1

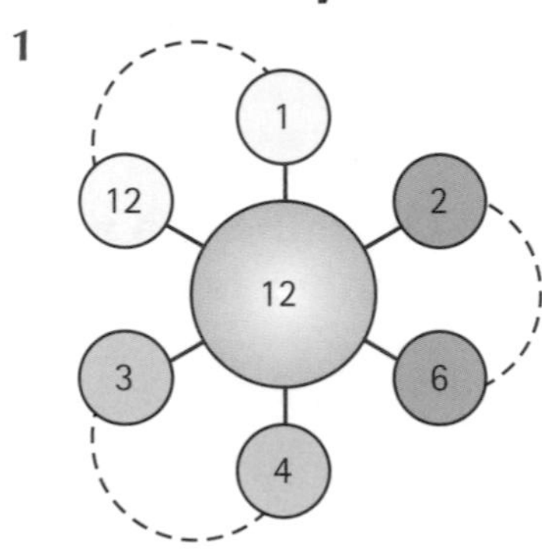

2

25, 1, 5, 25

3

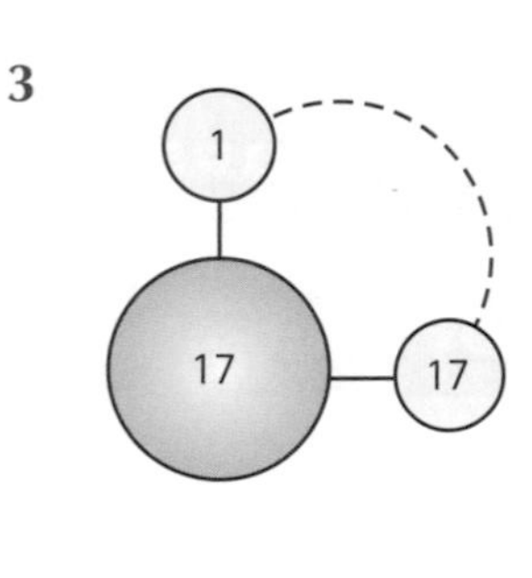

The factors of 12 are 1, 2, 3, 4, 6 and 12. They link up in **pairs**:

$1 \times 12 = 12$, $2 \times 6 = 12$ and $3 \times 4 = 12$.

2 25 has only three factors: 1, 5 and 25. 5 links with itself. Numbers like this are called **square numbers**.

3 Numbers that have only two factors are **prime**. They can only be divided exactly by 1 or themselves.

2 is the smallest prime number and is the only even one. Numbers that are not prime are **composite**.

4 Different numbers can share some of the same factors. These are called their **common factors**.

Factors of 12: ①, ②, ③, 4, ⑥, 12

Factors of 18: ①, ②, ③, ⑥, 9, 18

1, 2, 3 and 6 are the common factors of 12 and 18.

B Prime factors

1 Every whole number can be made by multiplying prime numbers together. Finding how to do this is called **decomposition.**

2 You can use a **factor tree** to decompose a number. Split every number into two factors. Branches that end in a prime number don't need to be split any further.

This tree shows that $90 = 2 \times 3 \times 3 \times 5$. Repeated prime factors can be written using indices: $2 \times 3^2 \times 5$ (see page 14).

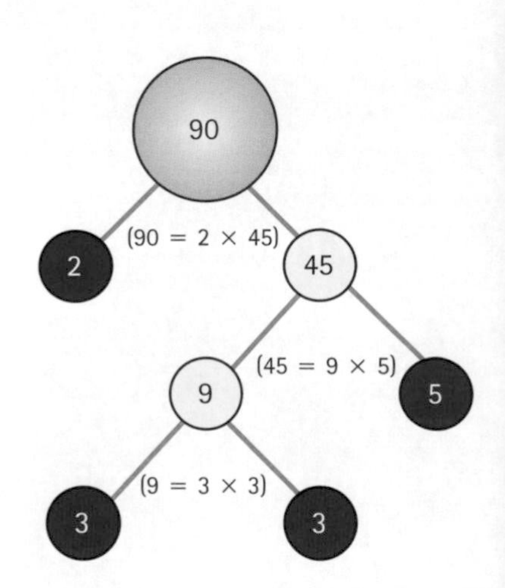

C Multiples

1 **key fact** **If you multiply a number by 1, 2, 3, 4, ... you calculate its multiples.**

The multiples of 12 are 12, 24, 36, 48, 60, ...

2 Different numbers can share some of the same multiples. These are called their **common multiples**.

Multiples of 12: 12, 24, (36), 48, 60, (72), 84, 96, (108), ...

Multiples of 18: 18, (36), 54, (72), 90, (108), ...

36, 72 and 108 are the first three common multiples of 12 and 18.

D HCF and LCM

1 **key fact** **HCF stands for highest common factor.**

For example, look at the common factors of 12 and 18 listed in section A4 on page 8. The largest is 6, so 6 is the HCF of 12 and 18.

2 You can use prime factors to find the HCF of two numbers.

Simply pick out the prime factors that are **common** to both numbers and multiply them together:

90 =	2 ×			3 ×	3 ×	5	
120 =	2 ×	2 ×	2 ×	3 ×		5	
HCF =	2 ×			3 ×		5	= 30

If you can't match anything, the HCF is 1.

3 **key fact** **LCM stands for lowest common multiple.**

Look at the common multiples of 12 and 18 listed above. The smallest is 36, so 36 is the LCM of 12 and 18.

4 You can use prime factors to find the LCM of two numbers.

Use the **biggest** power of each prime factor from the two numbers and multiply these together:

90 =	2 ×			3 ×	3 ×	5	
120 =	2 ×	2 ×	2 ×	3 ×		5	
LCM =	2 ×	2 ×	2 ×	3 ×	3 ×	5	= 360

>> practice questions

1 **List the factors of these numbers.**

(a) 36 (b) 27 (c) 50 (d) 30

2 **Use the lists you made in question 1 to write out the common factors of these pairs of numbers.**

(a) 30 and 36 (b) 27 and 30

(c) 50 and 30 (d) 50 and 27

3 **List the first ten multiples of these numbers.**

(a) 6 (b) 10 (c) 25 (d) 35

4 **Use the lists you made in question 3 to find some common multiples of these pairs of numbers.**

(a) 6 and 10 (b) 25 and 35

(c) 10 and 35 (d) 6 and 25

5 **Write these numbers as the product of their prime factors.**

(a) 36 (b) 42 (c) 40 (d) 63

6 **Use the lists you made in question 5 to find the HCF and LCM of these pairs of numbers.**

(a) 36 and 42 (b) 36 and 40

(c) 42 and 63 (d) 40 and 63

Working with fractions

- **The top of a fraction is called the numerator and the bottom is called the denominator.**
- **Fractions that contain different numbers but mean the same thing are equivalent. Given a fraction, the numerators and denominators of its equivalent fractions make simple sequences of multiples.**
- **The equivalent fraction with the smallest numbers is in 'lowest terms'.**

A Equivalent fractions

1 This diagram shows that $\frac{1}{3}$ is the same as $\frac{2}{6}$. It covers the same fraction of the diagram.

They are **equivalent fractions**.

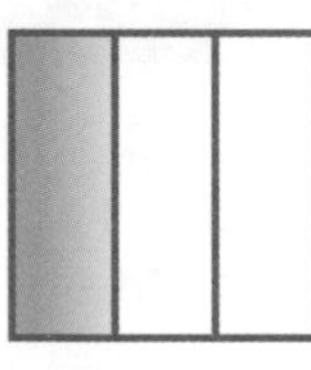

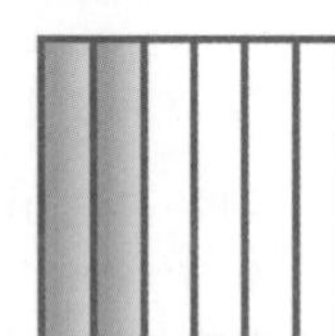

2 Look at this simple number pattern:

$\frac{1}{3} = \frac{2}{6} = \frac{3}{9} = \frac{4}{12}$ and so on.

These fractions are all equivalent.

They all equal $\frac{1}{3}$.

3 The fractions equivalent to $\frac{1}{3}$ contain numbers that make sequences.

$\frac{1}{3} = \frac{1\times 2}{3\times 2} = \frac{1\times 3}{3\times 3} = \frac{1\times 4}{3\times 4}$, etc.

B Improper fractions and mixed numbers

>> **key fact** **Improper fractions are top-heavy fractions. In other words they are fractions where the numerator is larger than the denominator, e.g. $\frac{5}{4}$.**

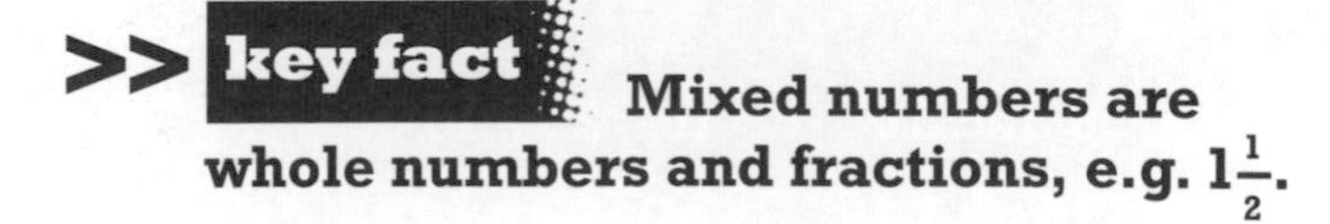

>> **key fact** **Mixed numbers are whole numbers and fractions, e.g. $1\frac{1}{2}$.**

1 Changing improper fractions to mixed numbers.

In the diagram, you can see $\frac{7}{3}$.

There are 2 wholes both made up of 3 thirds and a third left over, so that must mean that $\frac{7}{3} = 2\frac{1}{3}$.

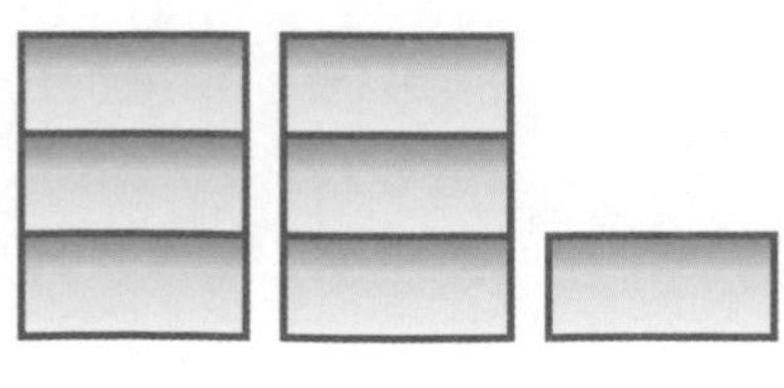

2 Changing mixed numbers into improper fractions.

The diagram shows $1\frac{2}{5}$. One whole $= \frac{5}{5}$.

$\frac{5}{5} + \frac{2}{5} = \frac{7}{5}$

The fraction part of the number is in fifths, so you need to convert it all to fifths.

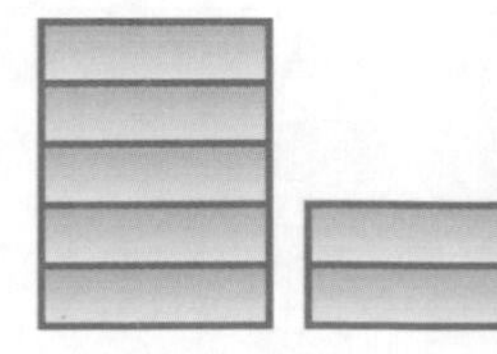

From the diagram, you can see that the whole block is split into $\frac{5}{5}$ and there are $\frac{2}{5}$ left over, so the answer is $\frac{7}{5}$.

C Adding and subtracting fractions

1 $\frac{1}{2} + \frac{1}{3} = ?$ You can see in this sum that the fractions have different denominators.

You need to make both denominators the same before you can add them.

2 $\frac{1}{2} = \frac{2}{4} = \frac{3}{6} = \frac{4}{8} = \frac{5}{10}$ (with $\frac{3}{6}$ circled)

$\frac{1}{3} = \frac{2}{6} = \frac{3}{9} = \frac{4}{12} = \frac{5}{15}$ (with $\frac{2}{6}$ circled)

The lowest denominator that is in both lists is 6.

3 Here we have changed them both to sixths: $\frac{1}{2} = \frac{3}{6}$ and $\frac{1}{3} = \frac{2}{6}$.

4 So $\frac{1}{2} + \frac{1}{3} = ?$ can be written as: $\frac{3}{6} + \frac{2}{6} = \frac{5}{6}$.

5 Subtraction works in the same way.

What is $\frac{5}{8} - \frac{7}{12}$?

The LCD for eighths and twelfths is 24, so change both fractions to 24ths.

$\frac{5}{8} = \frac{15}{24}$ and $\frac{7}{12} = \frac{14}{24}$.

$\frac{5}{8} - \frac{7}{12} = \frac{15}{24} - \frac{14}{24} = \frac{1}{24}$.

>> **key fact** **Given any two fractions, you can always find pairs of fractions with the same denominator that are equivalent to them. One of these pairs has the smallest denominator. This number is the lowest common denominator (LCD) of the two original fractions.**

D Multiplying and dividing fractions

1 To multiply two fractions together, simply multiply top by top and bottom by bottom.

$\frac{5}{8} \times \frac{4}{15} = \frac{5 \times 4}{8 \times 15} = \frac{20}{120}$. This cancels to $\frac{1}{6}$.

You can always get the answer this way, but it can sometimes lead to a lot of cancelling.

2 It can be easier to cross-cancel. This means doing some of the cancelling before you multiply.

3 Mixed numbers must be converted to improper fractions first.

To divide two fractions, invert the second one (turn it upside down) and turn the ÷ into a ×.

For example, $\frac{3}{4} \div \frac{3}{5} = \frac{3}{4} \times \frac{5}{3} = \frac{5}{4} = 1\frac{1}{4}$.

>> practice questions

1 Work out the equivalent fractions:

(a) $\frac{1}{2} = \frac{}{6}$ **(b)** $\frac{3}{4} = \frac{9}{}$ **(c)** $\frac{25}{40} = \frac{}{8}$ **(d)** $\frac{12}{27} = \frac{4}{}$

2 Change these improper fractions to mixed numbers:

(a) $\frac{15}{4}$ **(b)** $\frac{22}{7}$ **(c)** $\frac{51}{5}$ **(d)** $\frac{82}{9}$

3 Change these mixed numbers to improper fractions:

(a) $4\frac{1}{3}$ **(b)** $12\frac{1}{8}$ **(c)** $9\frac{7}{9}$ **(d)** $27\frac{3}{4}$

4 Calculate these:

(a) $\frac{2}{3} + \frac{1}{4}$ **(b)** $1\frac{1}{2} - \frac{3}{5}$ **(c)** $\frac{3}{4} \times \frac{8}{9}$ **(d)** $2\frac{1}{4} \div 1\frac{1}{2}$

Fractions, decimals and percentages

Any quantity can be written as a fraction, a decimal or a percentage.

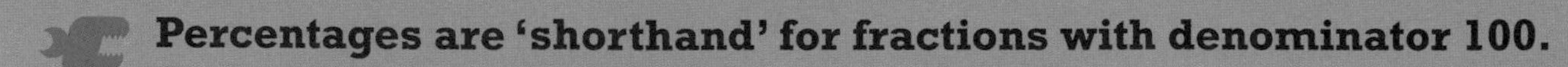

Percentages are 'shorthand' for fractions with denominator 100.

A Changing fractions to decimals

1 **key fact** **Remember that a fraction represents a division.**

So $\frac{2}{5}$ means $2 \div 5$.

$20 \div 5 = 4$, so $2 \div 5 = 0.4$
(see section B on page 2).

To change $\frac{5}{8}$ to a decimal, divide 5 by 8:

$$8\overline{)5.{}^{5}0{}^{2}0{}^{4}0}$$
$$0.6\ 2\ 5$$

2 Another way to change a fraction to a decimal is to find an equivalent fraction with 10, 100 or 1000 as the denominator. $\frac{2}{5} = \frac{4}{10}$ and $\frac{4}{10}$ equivalent to 0.4.

$\frac{5}{8} = \frac{625}{1000}$ (multiply top and bottom by 125)

$= 0.625$.

3 You should memorise these:

$\frac{1}{4}$	0.25	$\frac{1}{10}$	0.1	$\frac{1}{5}$	0.2	$\frac{1}{100}$	0.01
$\frac{1}{2}$	0.5	$\frac{3}{10}$	0.3	$\frac{2}{5}$	0.4	$\frac{1}{50}$	0.02
$\frac{3}{4}$	0.75	$\frac{7}{10}$	0.7	$\frac{3}{5}$	0.6	$\frac{1}{25}$	0.04
		$\frac{9}{10}$	0.9	$\frac{4}{5}$	0.8	$\frac{1}{20}$	0.05

4 Mixed numbers become decimals greater than 1.

$3\frac{5}{8} = 3.625$.

B Changing decimals to fractions

1 The decimal 0.36 means $\frac{3}{10} + \frac{6}{100} = \frac{30}{100} + \frac{6}{100} = \frac{36}{100}$.

Cancel to lowest terms to get $\frac{9}{25}$.

2 The quickest way to do this is to look at the number of digits after the decimal point.

If there's one, use 10 as denominator; if two, use 100, etc. Then use the digits as the numerator. So for 0.36, you know that 100 is the denominator and 36 the numerator.

3 Decimals greater than 1 become mixed numbers.

$2.36 = 2\frac{9}{25}$.

C Decimals and percentages

1 Decimals and percentages are very closely linked, so it's very easy to change from one to the other.

>> **key fact** **To change a decimal to a percentage, just multiply it by 100.**

The first two decimal places become the whole number part of the percentage and any other digits become the decimal part.

So 0.41 = 41%, 0.875 = 87.5% (or $87\frac{1}{2}$%), etc.

Be careful with decimals that have only one significant figure:

0.6 = 60%, 0.03 = 3%, etc.

2 **key fact** **To change a percentage to a decimal, divide by 100.**

So 25% = 0.25, 6.2% = 0.062, 70% = 0.7, etc.

3 Percentages over 100% become decimals larger than 1.

So 150% = 1.5, 200% = 2, etc.

D Recurring decimals

1 If the denominator of a fraction only has the prime factors 2 or 5, its decimal **terminates** (stops after a certain number of digits). Other denominators produce **recurring** decimals.

>> **key fact** **Recurring decimals contain a group of repeating digits that 'go on forever'.**

2 You can turn a fraction into its recurring decimal by dividing, as in section A on the previous page.

To change $\frac{5}{6}$ to a decimal, divide 5 by 6:

$$\begin{array}{r} 0.\,8\;3\;3\;3\;3 \\ 6\overline{)5.{}^{5}0{}^{2}0{}^{2}0{}^{2}0{}^{2}0} \end{array}$$

So $\frac{5}{6}$ = 0.833 33…, written $0.8\dot{3}$ to show that the 3 recurs (repeats).

If a group of digits recurs, put a dot over the first and last digit in the recurring part:

$\frac{1}{7} = 0.142\,857\,142\,857\,14\ldots = 0.\dot{1}42\,85\dot{7}$

You do **not** have to know how to change recurring decimals into fractions.

3 It is worth memorising these:

$\frac{1}{3} = 0.\dot{3}$, $\frac{2}{3} = 0.\dot{6}$, $\frac{1}{6} = 0.1\dot{6}$.

Note that your calculator will round recurring decimals: 1 ÷ 6 might be displayed as 0.166 666 67.

>> practice questions

1 Each column of the table represents one number and contains a fraction, decimal or percentage. Work out what should go in each empty cell. Write fractions in lowest terms.

	(a)	(b)	(c)	(d)	(e)	(f)	(g)	(h)	(i)	(j)	(k)	(l)
fraction		$\frac{9}{10}$		$\frac{1}{20}$			$\frac{3}{16}$				$\frac{1}{9}$	
decimal			0.35		0.44				2.375	0.072		
percentage	60%					12.5%		0.2%				145%

Powers and roots

- **Powers are a way of writing repeated multiplication. They consist of a base and an index.**
- **Negative indices produce reciprocals of the numbers produced by positive indices.**
- **To multiply or divide powers of the same base, add or subtract the indices.**

A Powers

1 Multiplication is a way of writing down repeated addition easily:

$3 + 3 + 3 + 3 + 3 + 3 = 3 \times 6.$

When you multiply repeatedly, you create a **power**.

$3 \times 3 \times 3 \times 3 \times 3 \times 3 = 3^6.$
This is pronounced 'three to the power six'.

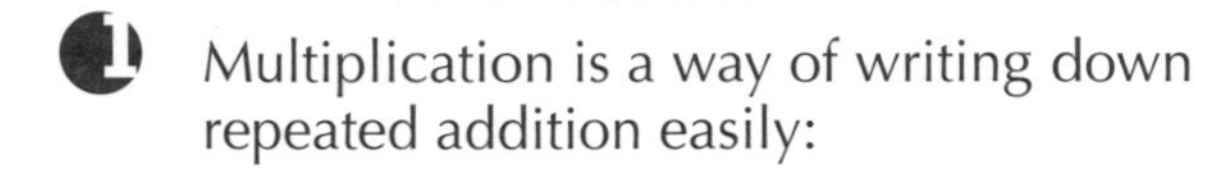

>> **key fact** **A power has two parts, a base and an index**

base → a^m ← index

The base is the number which is being multiplied and the index tells you how many 'copies' of the base to multiply together.

2 *Examples*: $5^2 = 5 \times 5 = 25$

$2^5 = 2 \times 2 \times 2 \times 2 \times 2 = 32$

Notice that 3^1 is just 3.

3 Some powers have special names.

3^2 is pronounced 'three squared'.

10^3 is pronounced 'ten cubed'.

4 Your calculator has keys for finding powers.

It has a key which will calculate any power $\square^{\blacksquare}$ or x^y,

a squaring key $\square^2$

and may even have a cubing key $\square^3$.

Make sure you know how to use them.

B Roots

1 **key fact** **Roots are the opposites or inverses of powers.**

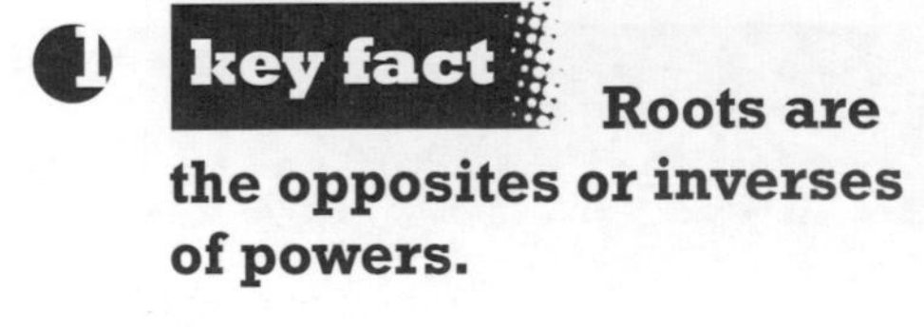

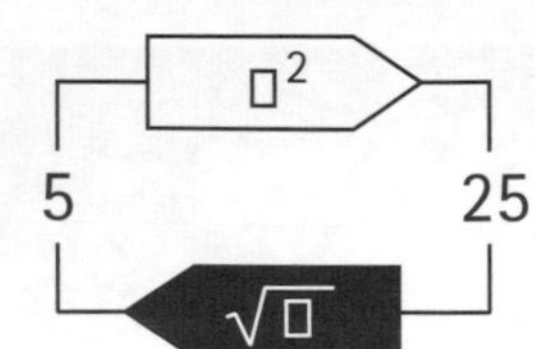

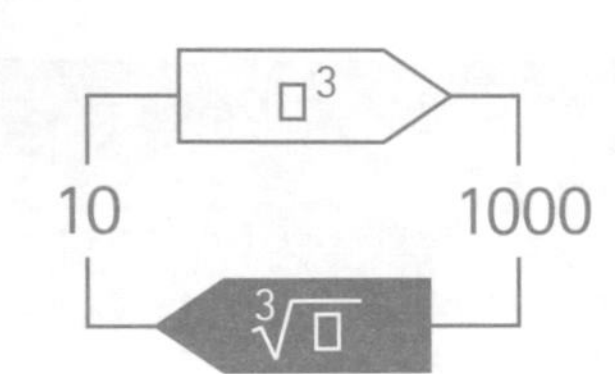

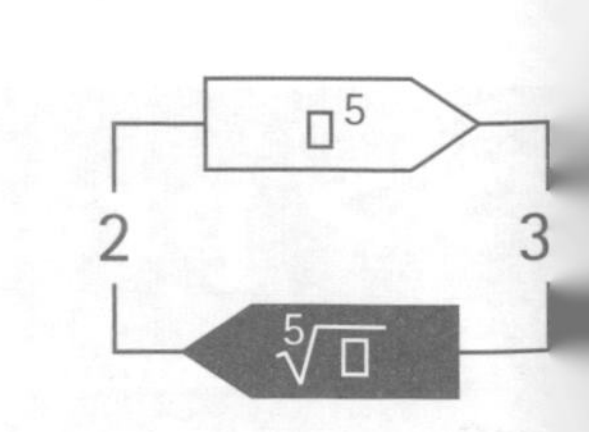

The diagrams above show that:

$\sqrt{25} = 5$, because $5^2 = 25$

$\sqrt[3]{1000} = 10$, because $10^3 = 1000$

$\sqrt[5]{32} = 2$, because $2^5 = 32$.

2 Your calculator has keys for finding roots.

It has a key which will calculate any root

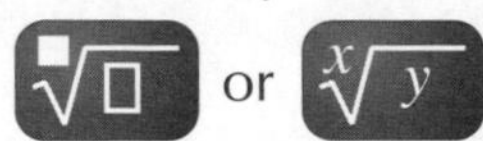

$\sqrt[\blacksquare]{\square}$ or $\sqrt[x]{y}$,

a square root key  $\sqrt{\square}$

and may even have a cube root key  $\sqrt[3]{\square}$.

C Zero and negative indices

1 In the table that follows, to move one column to the right, you divide by 2.

Power	2^3	2^2	2^1	2^0	2^{-1}	2^{-2}	2^{-3}
Value	8	4	2	1	$\frac{1}{2} = \frac{1}{2^1}$	$\frac{1}{4} = \frac{1}{2^2}$	$\frac{1}{8} = \frac{1}{2^3}$

This shows that >> **key fact** **Negative powers are the reciprocals of positive powers. Any number to the power zero is 1.**

So $3^{-2} = \frac{1}{3^2} = \frac{1}{9}$, $10^{-6} = \frac{1}{10^6} = \frac{1}{1\,000\,000}$, $7^{-3} = \frac{1}{7^3} = \frac{1}{343}$, etc.

D Combining powers

1 **key fact** **When two powers of the same base are multiplied together, add their indices.**

Example:

$5^2 \times 5^3 = (5 \times 5) \times (5 \times 5 \times 5) =$
$5 \times 5 \times 5 \times 5 \times 5 = 5^5$
so $5^2 \times 5^3 = 5^{(2 + 3)}$

Check:

$25 \times 125 = 3125$ ✓

2 **key fact** **When two powers of the same base are divided, subtract their indices.**

Example:

$10^6 \div 10^4 = 10^{(6 - 4)} = 10^2$

Check:

$1\,000\,000 \div 10\,000 = 100$ ✓

E The 'power of a power' rule

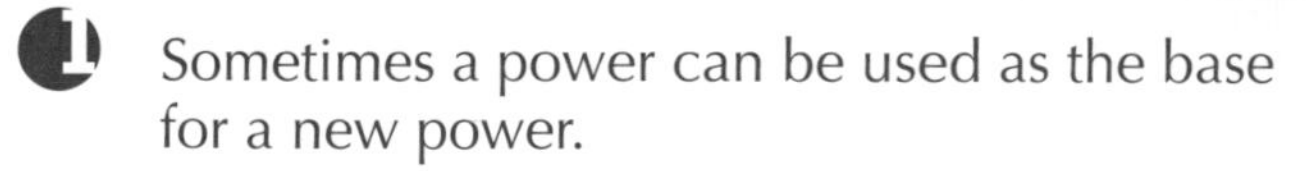

1 Sometimes a power can be used as the base for a new power.

$(3^2)^5 = 3^2 \times 3^2 \times 3^2 \times 3^2 \times 3^2 =$
$3^{(2+2+2+2+2)} = 3^{(2 \times 5)} = 3^{10}$.

This is known as the 'power of a power' rule. This is the only time that you can multiply indices.

2 You can use it to solve a problem like this:

Find x, if $9^2 = 3^x$

$9 = 3^2$, so $9^2 = (3^2)^2 = 3^{(2 \times 2)} = 3^4$

So $x = 4$

>> practice questions

1 **Calculate these powers and roots. Round answers to 3 significant figures if required.**

(a) 2^9 **(b)** 22^2 **(c)** 5^6 **(d)** 2^{-5}

(e) $\sqrt{64}$ **(f)** $\sqrt{7}$ **(g)** $\sqrt[5]{243}$ **(h)** $\sqrt[3]{10}$

2 **Use the index rules to find the missing numbers.**

(a) $4^3 \times 4^2 = 4^?$ **(b)** $10^6 \times 10^4 = 10^?$

(c) $5^7 \div 5^2 = 5^?$ **(d)** $2^4 \div 2^7 = 2^?$

(e) $(3^2)^4 = 3^?$ **(f)** $(1.5^3)^3 = 1.5^?$

dard index form

y number can be written in standard index form, but it is particularly useful for very large or very small numbers.

- **A number is in standard index form when it is written as $a \times 10^n$, where n is an integer and a is between 1 and 10.**
- **Large numbers have positive indices, small numbers (less than 1) have negative indices.**

A Powers of ten

1. Here are some of the powers of ten:

10^{-6}	10^{-3}	10^{-2}	10^{-1}	10^0	10^1	10^2	10^3	10^6
0.000 001	0.001	0.01	0.1	1	10	100	1000	1 000 000

These are used to write numbers in standard index form (also called *standard form* or *scientific notation*).

2. **key fact** **Numbers in standard index form consist of a number between 1 and 10, multiplied by a power of ten.**

3. Large numbers:

200 000	$= 2 \times 100000$	$= 2 \times 10^5$
4 250 000	$= 4.25 \times 1000000$	$= 4.25 \times 10^6$
560	$= 5.6 \times 100$	$= 5.6 \times 10^2$

4. Small numbers:

0.002	$= 2 \times 0.001$	$= 2 \times 10^{-3}$
0.000 073 3	$= 7.33 \times 0.00001$	$= 7.33 \times 10^{-5}$
0.89	$= 8.9 \times 0.1$	$= 8.9 \times 10^{-1}$

B Multiplication and division

1. To multiply two numbers in standard index form, rearrange them so the number and power parts can be dealt with separately (the brackets used here are just to make it easier to read).

$$(3 \times 10^4) \times (4 \times 10^7) = 3 \times 4 \times 10^4 \times 10^7$$
$$= 12 \times 10^{11}$$

This number is in index form, but not in *standard* index form, because the first part is bigger than 10. You need to adjust the size of the first number, then compensate with the power to keep the value the same:

12×10^{11}
÷ 10 × 10
1.2×10^{12}

2. To divide two numbers in standard index form, follow similar steps.

$$(1.2 \times 10^8) \div (2 \times 10^3) = 1.2 \div 2 \times 10^8 \div 10^3$$
$$= 0.6 \times 10^5$$

This also needs to be converted to standard index form.

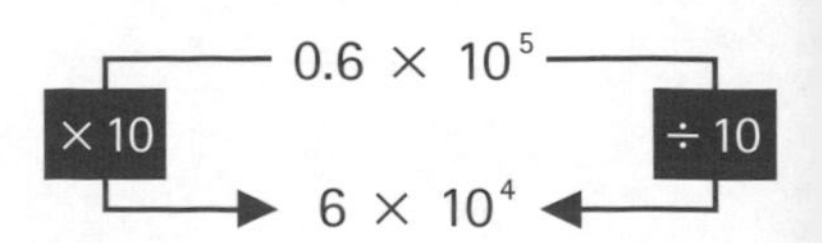

C Addition and subtraction

1 You can add or subtract numbers in standard index form, as long as the indices in the numbers are the same.

Examples: $(2 \times 10^5) + (4 \times 10^5) = 6 \times 10^5$

$(2.9 \times 10^9) - (2.4 \times 10^9) = 0.5 \times 10^9 = 5 \times 10^8$

2 If the indices are different, you need to adjust one of the numbers to match the other.

So $(8 \times 10^3) + (5 \times 10^4) = (0.8 \times 10^4) + (5 \times 10^4) = 5.8 \times 10^4$.

$(3 \times 10^{-6}) - (1.4 \times 10^{-7}) = (3 \times 10^{-6}) - (0.14 \times 10^{-6}) = 2.86 \times 10^{-6}$

D Using your calculator

1 Your calculator should have a key for entering numbers in standard index form.

It should look like one of these:

2 To enter 5×10^9, key in 5 EXP 9.

To enter 2×10^{-4}, key in

3 Once a number is entered, use it as you would a normal number. Try out some of the calculations from sections B and C to check that you're entering numbers correctly.

Never, under any circumstances, enter

This will give you 50×10^9, which is ten times too large!

Practice questions

1 Write these numbers in standard index form.

(a) 20 000 **(b)** 4 000 000

(c) 550 000 **(d)** 97 100 000 000

(e) 0.05 **(f)** 0.000 0003

(g) 0.000 72 **(h)** 0.355

2 Write these numbers in ordinary form (without indices).

(a) 6×10^7 **(b)** 3.23×10^5

(c) 1.9×10^{10} **(d)** 4×10^1

(e) 7×10^{-5} **(f)** 1.99×10^{-8}

(g) 9.03×10^{-2} **(h)** 8×10^{-10}

3 Work these out without using your calculator. Write your answers in standard index form.

(a) $(2 \times 10^5) \times (4 \times 10^2)$

(b) $(1.6 \times 10^7) \div (3.2 \times 10^5)$

(c) $(4 \times 10^4) + (9 \times 10^3)$

(d) $(5 \times 10^8) - (1.1 \times 10^7)$

4 Use your calculator to answer these. Write your answers in standard index form.

(a) $(1.5 \times 10^5)^2$

(b) $(3.3 \times 10^7) \times (2.1 \times 10^3) + (8 \times 10^8)$

(c) $\frac{(7.9 \times 10^5) - (3.3 \times 10^6)}{(2.5 \times 10^3)}$

(d) $\sqrt{(6.25 \times 10^{-8})}$

tio and proportion

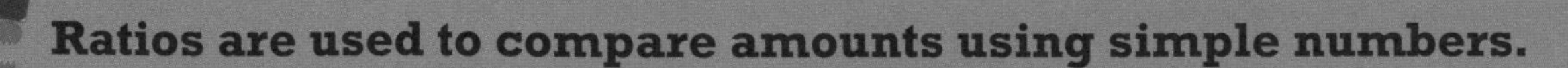
Ratios are used to compare amounts using simple numbers.

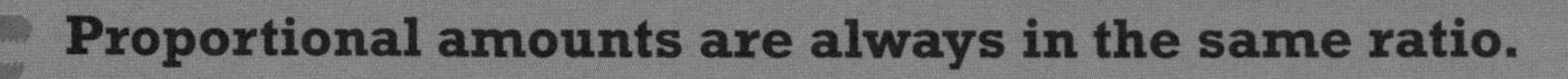
Proportional amounts are always in the same ratio.

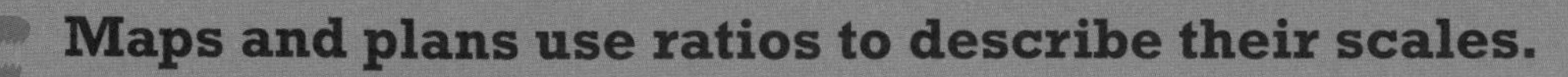
Maps and plans use ratios to describe their scales.

A Ratios

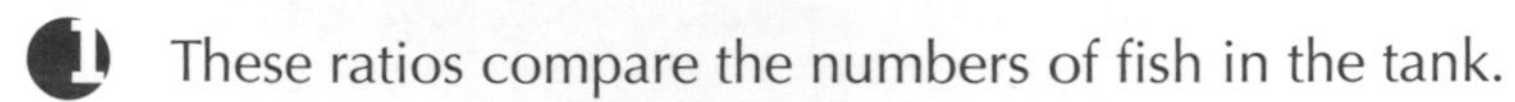
1. These ratios compare the numbers of fish in the tank.

There are 12 red fish and 8 green fish in the tank.
red : green = 12 : 8

For every 3 red fish, there are 2 green fish.
red : green = 3 : 2

2. **key fact** **Equivalent ratios, like equivalent fractions, contain different numbers but describe the same relationship.**

   You create equivalent ratios by multiplying or dividing all the numbers in a ratio by the same thing.

3. The equivalent ratio that contains the smallest whole numbers is in **lowest terms** or **simplest form**. 3 : 2 is the simplest form of 12 : 8.

4. A ratio that contains a 1 is called a **unitary** ratio. There are two unitary forms for 3 : 2.

   $3 : 2 = 1.5 : 1$ *divide both parts by 2*

   $3 : 2 = 1 : \frac{2}{3}$ *divide both parts by 3*

5. Ratios may have more than two parts. For example, in a recipe you might need 200 g flour, 100 g sugar and 50 g butter. The ratio is

   flour : sugar : butter = 200 : 100 : 50 = 4 : 2 : 1.

B Proportional division

1. You can divide up an amount according to a ratio. To do this:
   - Add up the numbers in the ratio. This tells you how many equal parts to divide the amount into.
   - Divide to find the amount in one part.
   - Multiply the amount in one part by each number in the ratio. These numbers solve the problem.

2. *Example*:

   Suppose £75 is to be divided between Sophie and her mother in the ratio 1 : 4. Follow the instructions: 1 + 4 = 5 parts.

   5 parts = £75
   1 part = £75 ÷ 5 = £15
   4 parts = £15 × 4 = £60

   So Sophie receives £15 and her mother £60.

C Proportional quantities

1 **key fact** **Two quantities are in direct proportion if they stay in a fixed ratio. For example, in these patterns, the ratio of circles to squares is always 2 : 1.**

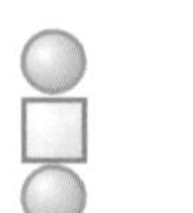 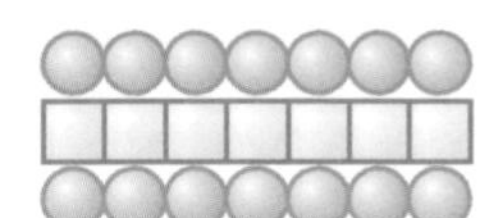 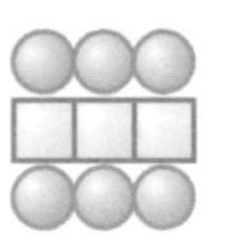 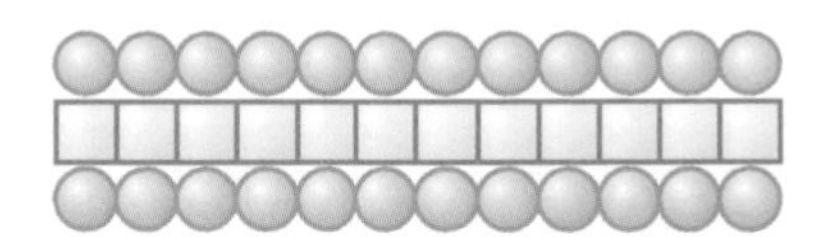

2 When solving problems involving proportional quantities, the **unitary method** is very useful. This involves changing the amount being varied to 1.

Example:

Suppose you can buy 5 litres of petrol for £4.00. How much would 12 litres cost?

5 litres	cost	£4.00	*write down what you know*
1 litre	costs	£4.00 ÷ 5 = £0.80	*change number of litres to 1*
12 litres	cost	£0.80 × 12 = **£9.60**	*multiply to obtain answer*

How much petrol could you buy for £15.00?

£4	will buy	5 litres	*write down what you know*
£1	will buy	5 ÷ 4 = 1.25 litres	*change number of £ to 1*
£15	will buy	1.25 × 15 = **18.75 litres**	*multiply to obtain answer*

3 Currency conversions involve proportional quantities. For example, 1 euro is worth about 90 pence, so euros : pounds = 1 : 0.9 = 1.11 : 1.

D Maps, scale models and plans

1 A popular scale for maps is 2 cm representing 1 km in reality.

The ratio of map sizes to actual sizes is
2 cm : 1 km = 2 cm : 1000 m =
2 cm : 100 000 cm.

This can now be written without units as
2 : 100 000 = 1 : 50 000.

It means that lengths on the map are $\frac{1}{50\,000}$ of the real lengths, or that real objects are 50 000 times larger than they appear on the map.

2 A scale model is also a fraction or multiple of the size of the real object. For example, if a model plane is made to a scale of 1 : 72 and is 15 cm long, the real plane is 72 × 15 = 1080 cm long, or 10.8 m.

3 The plan of a small electronic component might be 18 cm long, drawn at a scale of 40 : 1. In reality, its size is $\frac{1}{40}$ that of the drawing = 18 ÷ 40 = 0.45 cm.

>> practice questions

1 **Write these ratios in their simplest form.**
(a) 10 : 2 (b) 9 : 6
(c) 75 : 100 (d) 30 : 36

2 **Write these ratios in their unitary forms, 1 : *n* and *n* : 1.**
(a) 2 : 5 (b) 10 : 6

3 **Divide £150 in the ratios given.**
(a) 2 : 3 (b) 9 : 1

4 **100 ml of milk contain 50 calories; how many calories do 250 ml contain?**

5 **A ship that is 120 m long is modelled on a scale of 1 : 250. How long is the model?**

Percentage calculations

- **To calculate a percentage divide by 100 to find 1%, then multiply by the number of %.**
- **Percentage increases or decreases can be added to or subtracted from the original amount, or the new percentage calculated.**

A Percentages of an amount

1. **key fact** **Remember that percentages are just fractions 'out of 100'.**

You can use the **unitary method** to calculate a percentage of something, by finding 1% first.

Example: What is 6% of £350?

100% = £350
1% = £350 ÷ 100 = £3.50 *divide by 100 to find 1%*
6% = £3.50 × 6 = £21 *multiply by 6 to get required number of %*

2. If you know that the percentage you want is the same as a simple fraction, you can use this to make the calculation easier.

Example: Find 25% of £60.

25% is the same as $\frac{1}{4}$, so 25% of £60 = £60 ÷ 4 = £15.

3. Calculations like that in 1 above can be done 'in one go' on your calculator.

To find $17\frac{1}{2}$% of £26, you would enter

1 7 . 5 ÷ 1 0 0 × 2 6 =

Alternatively, you could change $17\frac{1}{2}$% to a decimal, 0.175, and just multiply £26 by 0.175.

B Percentage increases

1. **key fact** **Percentage increases occur when things grow or sums of money have interest added.**

There are two ways to calculate a percentage increase.

Suppose you could buy 10 litres of petrol for £8, but this is increased by 5%. What is the new price?

2. *Method 1* – Calculate the increase, then add this onto the original price.

5% of £8 = £8 ÷ 100 × 5 = £0.40.

New price = £8 + £0.40 = £8.40.

3. *Method 2* – Find the new total percentage, then calculate this.

100% + 5% = 105%.

105% of £8 = £8 × 1.05 = £8.40.

C Percentage decreases

1 key fact Percentage decreases occur when things shrink, or sums of money have tax deducted or are discounted.

The two methods above apply to percentage decreases.

Suppose a pair of jeans costing £26 is reduced by 15% in a sale. What is the discounted price?

2 *Method 1* – Calculate the decrease, then subtract this from the original price.

15% of £26 $= £26 \div 100 \times 15 = £3.90$.

Sale price $= £26 - £3.90 = £22.10$.

3 *Method 2* – Work out what percentage of the original price is being charged.

$100\% - 15\% = 85\%$.

85% of £26 $= £26 \times 0.85 = £22.10$.

D Finding the percentage

1 key fact Expressing one amount as a percentage of another is the same as converting a fraction to a percentage.

There are two ways to approach this kind of problem.

Suppose you scored 24 out of 30 in a test. What percentage is this?

2 *Method 1* – Divide one amount by the other, then multiply by 100.

$24 \div 30 \times 100 = 80\%$

Method 2 – Write the division as a fraction and find an equivalent fraction with denominator 100.

$$\frac{24}{30} = \frac{8}{10} = \frac{80}{100} = 80\%$$

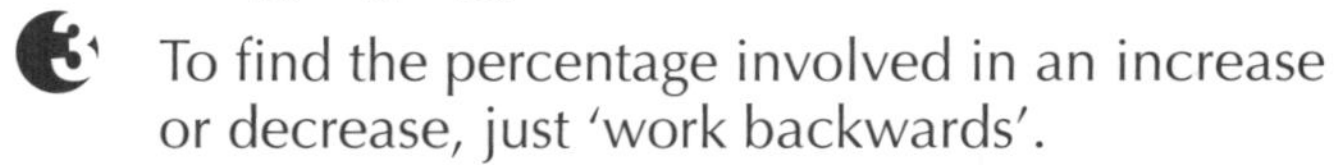

3 To find the percentage involved in an increase or decrease, just 'work backwards'.

Example: A culture of 20 000 bacteria increases to 27 000. What is the percentage growth?

Increase $= 27\,000 - 20\,000 = 7000$.

Now find this as a percentage of the original amount.

Percentage increase
$= 7000 \div 20\,000 \times 100 = 35\%$.

>> practice questions

1 Find these percentages.

	£40	£22.50	15 cm	3000 people
25% of ...				
7% of ...				

2 Calculate the increases and decreases.

	£200	£11	25 litres	500 tonnes
Increase by 3% ...				
Decrease by 20% ...				

3 Convert the following to percentages.

(a) 28 out of 40 in a test. **(b) 4500 people out of 30 000.**

4 Calculate the percentage increase or decrease.

(a) £300 changed to £360. **(b) 500 people changed to 370.**

Algebraic expressions

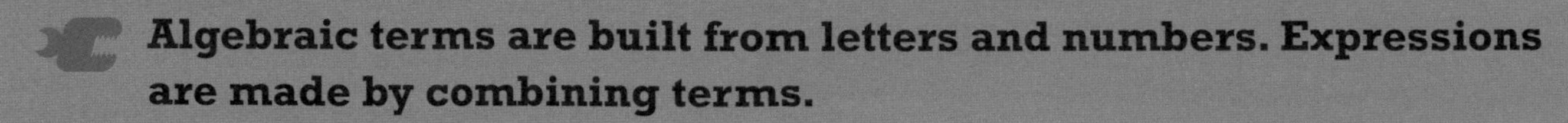

Algebraic terms are built from letters and numbers. Expressions are made by combining terms.

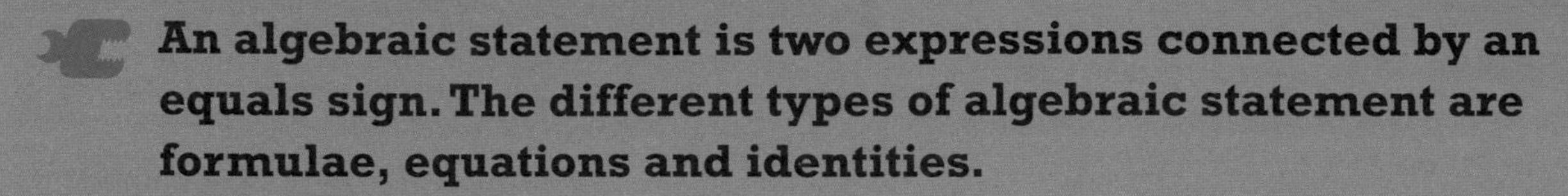

An algebraic statement is two expressions connected by an equals sign. The different types of algebraic statement are formulae, equations and identities.

A Notation

1 In algebra, letters are used to stand for numbers. A letter may stand for an unknown number, or it may indicate a place where you can use any number you wish.

>> **key fact** **Letters represent variables or unknowns and are usually printed in an *italic* font.**

2 The standard operations that you can use on numbers apply to letters in algebra too. However, there are some special ways of writing them down.

Multiplication:

$2x$ means '2 times the number x'.
ab means 'multiply the numbers a and b together'.

Division: $\frac{z}{10}$ means 'divide the number z by 10'.

B Terms

1 A **term** in algebra can be made up from:

- single letters or numbers
- letters and numbers that are linked by multiplication or division.

2 *Examples*:

$x \quad a \quad 1 \quad 25 \quad 2x \quad ab \quad 5x^2 \quad \frac{N}{2} \quad \frac{24xyz}{5ac}$

C Expressions

1 An **expression** in algebra can be made up from:

- a single term
- a number of terms linked by adding or subtracting.

Examples:

$x + c \qquad n - 5 \qquad 2xy + 2zy + 2zx$

$\frac{p^2}{2} + \frac{q^2}{3} - \frac{t^2}{10}$

2 You can build up more complicated expressions by multiplying or dividing simpler expressions.

Examples:

$(x + 1)(2x - 3) \qquad \frac{2d+5}{d+5}$

D Formulae

1 **key fact** **A formula is a set of instructions for calculating something.**

The thing being calculated is called the **subject** of the formula and is usually a single letter on the left of the equals sign. The instructions appear as an expression on the right of the equals sign.

2 *Examples*:

- To calculate the perimeter of a rectangle (P) if you know its length (l) and width (w), double the length, double the width and add these together.

  The formula is $P = 2l + 2w$.

- To calculate the average speed of a journey (s) if you know the distance (d) and time taken (t), divide the distance by the time.

  The formula is $s = \frac{d}{t}$.

E Equations

1 **key fact** **An equation is like a mathematical puzzle which requires a solution.**

A letter (often x) stands for the unknown number in the equation. You solve the equation by simplifying it according to the rules of algebra.

2 *Examples*: $2x + 3 = 11$ (solution: $x = 4$)

$\frac{x}{5} = 20$ (solution: $x = 100$)

3 Formulae can sometimes produce equations to solve. Suppose a rectangle's perimeter is 32 cm and its width is 5 cm. Use the formula for the perimeter from section D.

$P = 2l + 2w$. Substitute the known values.

$32 = 2l + 10$. This is now an equation to find l.

F Identities

1 **key fact** **An identity is a mathematical statement that is true whatever values the letters take. Identities often show you two different ways of writing the same thing.**

2 *Example*: You can calculate the perimeter of a rectangle like this: add the width to the length, then double the result.
This would be written $P = 2(l + w)$.

So $2(l + w) = 2l + 2w$. This is an identity, because it is true whatever values you choose for l and w.

>> practice questions

1 How many terms are there in each expression?

(a) $3x + 5$ **(b)** ab **(c)** $P + Q - R$ **(d)** 1357 **(e)** $\frac{u}{v}$

(f) c^2 **(g)** $x^2 + 2xy + y^2$ **(h)** πr^2 **(i)** $abcdef$ **(j)** $t^3 - 3t^2 + 3t - 1$

2 Decide whether each of these algebraic statements is a formula, an equation or an identity.

(a) $3p + 5 = 20$ **(b)** $2x + 2x = 4x$ **(c)** $12 = 10 - x$ **(d)** $C = \pi d$

(e) $x^2 + 5x = 0$ **(f)** $F = \frac{9C}{5} + 32$ **(g)** $(x + y)(x - y) = x^2 - y^2$ **(h)** $z + 6 = 2z + 8$

(i) $c^2 = a^2 + b^2$ **(j)** $(pq)^2 = p^2q^2$

Formulae and substitution

The subject of a formula is the thing the formula is designed to calculate.

Substituting numbers into a formula replaces letters by their values.

A The subject of a formula

1. **key fact** **A formula is a set of instructions for calculating something.**

The thing being calculated is called the **subject** of the formula and is usually a single letter on the left of the equals sign. The instructions appear as an expression on the right of the equals sign.

2. *Example*:

- To calculate the area of a triangle (A) if you know the length of its base (b) and its perpendicular height (h), multiply them together, then halve the result.

The formula is $A = \frac{bh}{2}$.

B Substitution

1. To use a formula to carry out the calculation it describes, you need to **substitute** numbers for letters in the formula.

>> **key fact** **Substitution means replacing each known letter in a formula by its value.**

2. Use the word VASE to help you remember how to set things out:

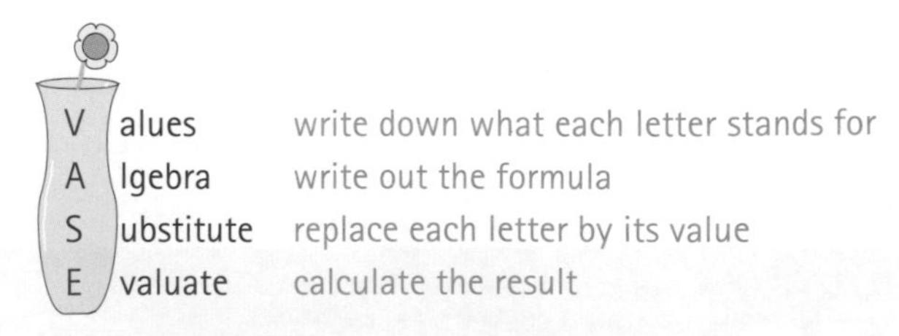

V alues	write down what each letter stands for
A lgebra	write out the formula
S ubstitute	replace each letter by its value
E valuate	calculate the result

Example:

The perimeter of a triangle, P cm, whose sides are a cm, b cm and c cm long, is

$$P = a + b + c.$$

Find the perimeter if $a = 4$, $b = 10$ and $c = 7$.

V	$a = 4,\ b = 10,\ c = 7$	
A	P	$= a + b + c$
S		$= 4 + 10 + 7$
E		$= \underline{21}$

3. Be careful when the formula contains multiplication.

For example, if $y = 2x$ and $x = 3$, when you substitute you must remember to put back the 'invisible' multiplication sign: $y = 2 \times 3$, not $y = 23$!

4. Take extra care if any of the numbers you need to substitute are negative. Make sure you obey the rules for working with negative numbers.

Example:

Find the value of R using the formula $R = 5x - 3y$, if $x = 2$ and $y = -4$.

V	$x = 2,\ y = -4$	
A	R	$= 5x - 3y$
S		$= 5 \times 2 - 3 \times -4$
E		$= 10 - (-12)$
		$= 10 + 12$
		$= \underline{22}$

C Creating formulae

1. A formula is a bit like a mathematical recipe.
You need ingredients (the letters and numbers used in the formula) and a method (a way to combine the ingredients).

2. *Example*:

To calculate a Fahrenheit temperature from a Celsius temperature:

'Ingredients': Celsius temperature (C), Fahrenheit temperature (F).

'Method': Multiply the Celsius temperature by 9, then divide by 5. Add 32 to the result.

The formula is $F = \frac{9C}{5} + 32$.

3. Always check any formula you create by substituting some suitable numbers into it. In the formula above, you could use the fact that the boiling point of water is 212°F or 100°C.

V $F = 212,\ C = 100$

A $F = \frac{9C}{5} + 32$

S $212 = \frac{9 \times 100}{5} + 32$

E $= \frac{900}{5} + 32$

$= 180 + 32$

$= 212$ ✓

Practice questions

1 **Given that $a = 2$, $b = 3$, $c = 10$, $x = 0.1$ and $y = -4$, evaluate the following expressions.**

(a) $a + b + c$ (b) $5c - 8$

(c) $ab + y$ (d) $\frac{x}{c}$

(e) $\frac{12a - 7b}{cx}$ (f) c^2

(g) $4ay$ (h) $(cy)^2$

(i) $5(ac + by)$ (j) $\sqrt{bc + 3a}$

2 **Use the formula $Z = \frac{u - v}{u + w}$ to calculate Z for each of these sets of values.**

(a) $u = 7$, $v = 1$, $w = 3$

(b) $u = 0.75$, $v = 0.57$, $w = 0.15$

(c) $u = 32$, $v = 8$, $w = 96$

(d) $u = 3500$, $v = 500$, $w = -2500$

(e) $u = 2.12$, $v = 0.62$, $w = 2.88$

3 **Use the given instructions to create formulae.**

(a) To calculate your Body Mass Index (B) you need to know your height in metres (h) and your weight in kilograms (w). Square your height, then divide your weight by the result.

(b) To work out the surface area (S) of a cuboid, you need its width (w), length (l) and height (h). Multiply the length by the width, then multiply the width by the height, then the height by the length. Add the three results together, then double the answer.

Solving equations

- **To rearrange an equation, add the same number to both sides, subtract the same number from both sides, or multiply or divide both sides by the same number.**
- **Whatever you do to one side of the equation, you must do exactly the same to the other side.**
- **To 'undo' an operation, perform the opposite or inverse operation.**

A Solving equations with an unknown on one side

>> **key fact** **An equation is like a puzzle. Finding the correct value for the letter in an equation solves the puzzle. This value is the solution to the equation.**

1. You can solve this equation: $3x + 4 = 19$

$3x + 4 - 4 = 19 - 4$ *first take 4 from both sides*

$3x = 15$

So $x = 5$ *now divide both sides by 3*

2. Solve $5x - 1 = 34$

$5x - 1 + 1 = 34 + 1$ *add 1 to both sides*

$5x = 35$

So $x = 7$ *divide both sides of the equation by 5*

B Solving equations with unknowns on both sides

1. Look at this equation: $4x - 6 = 3x + 3$

Step 1: Remove the x term from the right-hand side.

$4x - 6 - 3x = 3x + 3 - 3x$ *subtract 3x from both sides*

$x - 6 = 3$

Step 2: Remove the number term from the left-hand side.

$x - 6 + 6 = 3 + 6$ *add 6 to both sides*

$x = 9$

C Where the x term contains a fraction

There are two cases:
either the unknown is the denominator of the fraction or the numerator.

$\frac{x}{5} - 7 = 10$

$\frac{x}{5} = 17$ *add 7 to both sides*

$x = 85$ *multiply both sides by 5*

$\frac{2}{x} + 3 = 7$

$\frac{2}{x} = 4$ *subtract 3 from both sides*

$2 = 4x$ *multiply both sides by x*

$0.5 = x$ *divide both sides by 4*

D Equations with a negative x term

1 This equation has a negative unknown on one side: $5 - 6x = 3$

$5 - 6x + 6x = 3 + 6x$ *add 6x to both sides*

$5 = 3 + 6x$

$5 - 3 = 3 + 6x - 3$ *subtract 3 from both sides*

$2 = 6x$

So $x = \frac{1}{3}$

2 Solve $5x + 2 = 1 - x$

$5x + 2 + x = 1 - x + x$ *add x to both sides*

$6x + 2 = 1$

$6x + 2 - 2 = 1 - 2$ *subtract 2 from both sides*

$6x = -1$

$x = -\frac{1}{6}$

>> practice questions

Solve these equations and check your answers.

1 $3x + 2 = 23$

2 $5y + 7 = 52$

3 $8d + 3 = 19$

4 $10f + 7 = 127$

5 $6c - 1 = 5c + 4$

6 $7h - 3 = 11h - 5$

7 $4x + 5 = 10x - 13$

8 $3x + 7 = 3 - x$

9 $9 - \frac{1}{3}k = 2k - 5$

10 $4 - 8p = 2 - 5p$

Trial and improvement

- Sometimes you can't find an exact solution to an equation, but can find a reasonable **approximation** using **trial and improvement**.
- The accuracy of the approximation depends on the number of trials: the more trials you do, the closer you can get to the solution.

A Decimal searches

1 *Example*:

Solve the equation $x^2 = 3$, correct to 2 decimal places.

This is a very simple equation which you could solve easily with a press of the square root key! However, it demonstrates the solution process clearly. Notice how the search concentrates on whole numbers, then decimals with one decimal place, then two.

>> **key fact** **In a decimal search, the solution is found one significant figure at a time.**

2 Record the results of the trials, together with your decisions, in a table like the one opposite:

x	x^2	Comments
1	1	Too small, so $x > 1$. Try $x = 2$.
2	4	Too big, so $x < 2$. x is between 1 and 2. Move on to numbers with 1 decimal place: try 1.5 first, as it's halfway between 1 and 2.
1.5	2.25	Too small, so $x > 1.5$. Try 1.7, as it's about halfway between 1.5 and 2.
1.7	2.89	Too small, so $x > 1.7$. Try 1.8.
1.8	3.24	Too big, so $x < 1.8$. x is between 1.7 and 1.8. Move on to numbers with 2 decimal places. Try 1.75, as it's halfway between 1.7 and 1.8.
1.75	3.0625	Too big, so $x < 1.75$. Try 1.72, as it's about halfway between 1.7 and 1.75.
1.72	2.9584	Too small, so $x > 1.72$. Try 1.73.
1.73	2.9929	Too small, so $x > 1.73$. Try 1.74.
1.74	3.0276	Too big, so $x < 1.74$. x is between 1.73 and 1.74, so the answer is one of these. You now need to know whether it's closer to 1.73 or 1.74. Trying 1.735 will decide.
1.735	3.010225	Too big, so $x < 1.735$. x is between 1.73 and 1.735.

Therefore the solution is $x = 1.73$, to 2 dp.

3 You have to test $x = 1.735$, because you need to know whether x is closer to 1.73 or 1.74. It's closer to 1.73, so whatever the *exact* solution of the equation is, when rounded to 2 dp, it's 1.73.

>> **key fact** **You try out likely solutions in an equation to see how closely they fit. You use the results to make better guesses. That's why it's 'trial and *improvement*', not just 'trial and *error*'.**

B Rearranging the equation

1 *Example*:

Solve the equation $p^3 - 10 = p$, correct to 1 decimal place.

As you try different values for p, both sides of the equation change. It's much easier to 'hit the target' if you're aiming for a fixed number, so rearrange the equation to read $p^3 - p = 10$.

p	$p^3 - p$	Comments
1	0	Too small, so $p > 1$. Try $p = 2$.
2	6	Too small, so $p > 2$. Try $p = 3$.
3	24	Too big, so $p < 3$. p is between 2 and 3. Move on to numbers with 1 decimal place: try 2.5 first, as it's halfway between 2 and 3.
2.5	13.125	Too big, so $p < 2.5$. Try 2.3, as it's about halfway between 2 and 2.5.
2.3	9.867	Too small, so $p > 2.3$. Try 2.4.
2.4	11.424	Too big, so $p < 2.4$. p is between 2.3 and 2.4, so the answer is one of these. You now need to know whether it's closer to 2.3 or 2.4. Trying 2.35 will decide.
2.35	10.627875	Too big, so $p < 2.35$. p is between 2.3 and 2.35.

Therefore the solution is $p = 2.3$, to 1 dp.

As you might have guessed, the solution ($p = 2.3089...$) is very close to 2.3!

C Exact answers

1 Sometimes you may locate an exact answer by trial and improvement. Suppose you had to solve

$$\frac{552.96}{x^3} = 5.$$

By trying whole numbers, you would find that x is between 4 and 5.

Looking between 4 and 5, you would eventually try $x = 4.8$. This is exactly right, so no rounding of the answer is necessary.

>> practice questions

Use trial and improvement to find solutions to the following equations. Give answers correct to 1 decimal place.

1 $x^3 - 2x^2 = 1$
2 $x^3 + 4x - 18 = 0$
3 $5y^2 - 17y = 1.75$
4 $6m^2 - 9 = 6m$
5 $5t + 1.5 = 2t^3$
6 $5x^3 = 60.835$

Rearranging formulae

- **Rearranging formulae means getting the subject on one side of the equals sign and everything else on the other side of the equals sign.**
- **It is also known as changing the subject, or transforming formulae.**

A Changing the subject – types 1 and 2

>> **key fact** **Changing the subject of a formula means rearranging a formula to get one letter on its own and all of the other letters on to the other side of the equals sign.**

1. It is also known as **transformation** of formulae.

Type 1: When x is not 'bound' up with anything else:

$x + a = b$

$x + a - a = b - a$ *here subtract a from both sides*

$x = b - a$

Type 2: When x is 'bound' in a multiplication:

$xy = a$

$\frac{xy}{y} = \frac{a}{y}$ *divide both sides by y*

$x = \frac{a}{y}$

2. Look at this formula: $2x^2y = z$

$x^2 = \frac{z}{2y}$ *divide both sides by 2y*

$x = \sqrt{\frac{z}{2y}}$ *remove the square by finding the square root of both sides*

B Combinations of types 1 and 2

1. Make x the subject of this formula:

$q + 6x = p$

$q + 6x - q = p - q$ *isolate the xs first by taking q from both sides*

$6x = p - q$

2. We need only one x on the left-hand side, so we 'undo' $6x$ by dividing each side by 6:

$x = \frac{p-q}{6}$

C What to do if the subject has a minus sign

1. Sometimes you may arrange a formula only to find that the left-hand side contains the subject with a minus sign in front of it, for example $-A = sx - 2t$.

>> **key fact** **If this happens, multiply both sides of the equation by -1.**

This has the effect of changing the sign of **every term** in the formula, because: $-A \times -1 = A$.

If $-A = sx - 2t$, then $A = -sx + 2t$

2. Make r the subject of $3M - 2r = 4N$.

$-2r = 4N - 3M$ *subtract 3M from both sides*

$-r = 2N - \frac{3M}{2}$ *divide both sides by 2*

$r = -2N + \frac{3M}{2}$ *multiply both sides by −1*

$r = \frac{3M}{2} - 2N$ *rearrange right-hand side for neatness*

D When factorisation is required

1. If the letter you want to make the subject of a formula occurs in **two terms**, you may need to **collect** these terms and then **factorise** them.

2. Make x the subject of $2x = px + q$.

$2x - px = q$ *subtract px from both sides*

$(2 - p)x = q$ *factorise the left hand side*

$x = \frac{q}{2-p}$ *divide both sides by 2 − p*

>> practice questions

In these questions, make x the subject of the formula.

1 $x + 9 = r$

2 $x - z = a$

3 $5x + 4y = 16$

4 $mx^2 = c$

5 $\frac{1}{4}x = m$

6 $\frac{x}{p} = p + c$

7 $\frac{mx}{b} = c$

8 $5 - fx = 3x + p$

In these questions, make y the subject of the formula.

9 $2x + 5y = 9$

10 $x - 2y = 10$

Using brackets in algebra

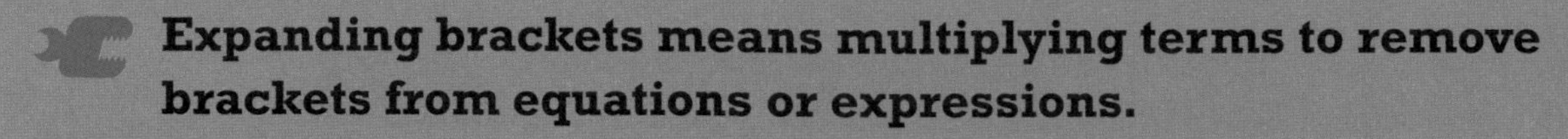

Expanding brackets means multiplying terms to remove brackets from equations or expressions.

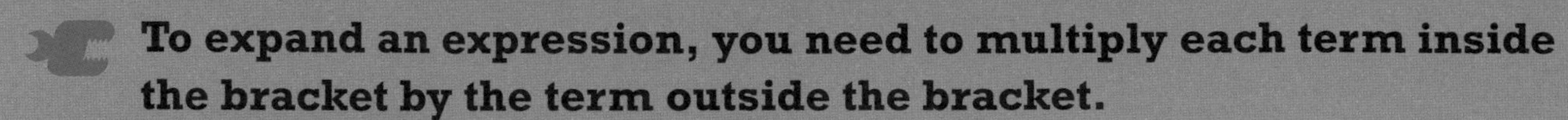

To expand an expression, you need to multiply each term inside the bracket by the term outside the bracket.

A Expanding brackets

>> **key fact** **To expand brackets, multiply everything in the bracket by the term outside the bracket.**

1. Expand $3(2a + b) \quad = 6a + 3b$
2. Expand $6(y + 3) \quad = 6y + 18$

   It is a common mistake to forget to multiply the **second** term.
   If you worked out $6y + 3$ for the answer to the second equation, you'd be wrong!
3. Expand $3(2x + 4) = 6x + 12$

   This example uses the fact that $3 \times 2x = 6x$.
4. Expand $x(4x + 9) = 4x^2 + 9x$

   Notice here that x is the term outside the bracket and that $x \times x = x^2$.
5. Expand $5x(3x - 2) = 15x^2 - 10x$

   Notice that $5x \times 3x = 5 \times 3 \times x \times x = 15x^2$.

B Factorising expressions

>> **key fact** **Factorising is the opposite of expanding brackets. Find the highest common factor (HCF) of all of the terms in the expression you are trying to factorise. The HCF must appear outside the brackets.**

Factorise $4x^2 + 8x$.

1. Here the HCF is the largest term that is a factor of $4x^2$ and $8x$. The HCF must be $4x$.
2. So we now have $4x^2 + 8x = 4x(? + ?)$.
3. Ask yourself, 'What do I multiply $4x$ by, to make $4x^2$, and what do I multiply $4x$ by to make $8x$?'
4. So the final answer $= 4x(x + 2)$.

C Solving equations with brackets

Method 1

1. Solve $4(x + 3) = 28$.
2. Expand the brackets:

$$4x + 12 = 28$$

3. Solve it like any other equation:

$$4x + 12 = 28$$
$$4x = 16$$
$$x = 4$$

Method 2

1. Solve $4(x + 3) = 28$.
2. $4(x + 3)$ is a product, it is $4 \times (x + 3)$.
3. So divide by 4 to undo the multiplication:

$$x + 3 = 7$$
$$x = 4$$

D Solving equations with two sets of brackets

Solve $3(x + 4) + 5(x - 6) = 5x - 3$.

Step 1: Expand the brackets and simplify.

$$3x + 12 + 5x - 30 = 5x - 3$$
$$8x - 18 = 5x - 3$$

Step 2: Solve the equation. You should find that:

$$x = 5$$

>> practice questions

1 $2(x + 5) = 18$

2 $5(x - 2) = 40$

3 $4(3x - 7) - 5(2x - 4) = 3x$

4 $2(7x - 7) - 2(4x + 5) = 21 - 4x$

5 $5 + (7x - 7) - 2(4x + 5) = 2x + 4$

6 $6(4x - 7) + 3(13 - 3x) = 20x - 23$

7 $8(2x - 9) - 2(9 - 2x) = 14x + 3$

8 **Factorise:**

(a) $14x^2 + 7x$ **(b)** $36y^2 - 9y$

9 **Factorise:**

(a) $15y^4 + 25y^2$ **(b)** $100a^2 + 20ab^3$

Multiplying bracketed expressions

You need to know how to expand expressions made by multiplying two brackets, such as $(w + x)(y + z)$.

Both terms in the first bracket have to be multiplied by both terms in the second, resulting in four new terms.

A Bracketed expressions

1. Look at this rectangle.

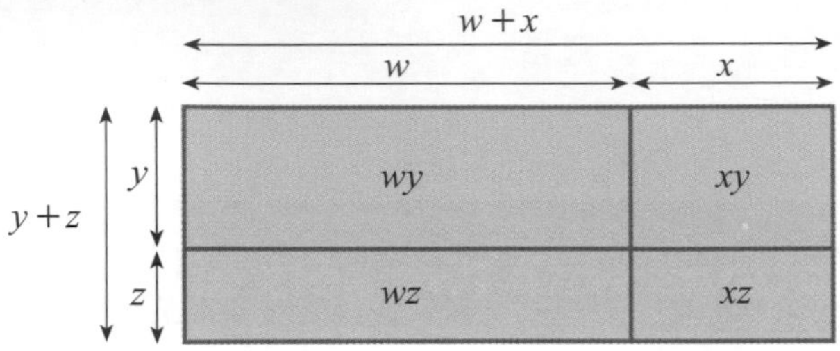

	w	x
y	wy	xy
z	wz	xz

(Total width $w + x$; total height $y + z$.)

2. The area of this rectangle is $(w + x)(y + z)$.
3. We can write this as the combined area of the four smaller rectangles:

$wy + wz + xy + xz$

B More difficult expansions

1. Expand and simplify:

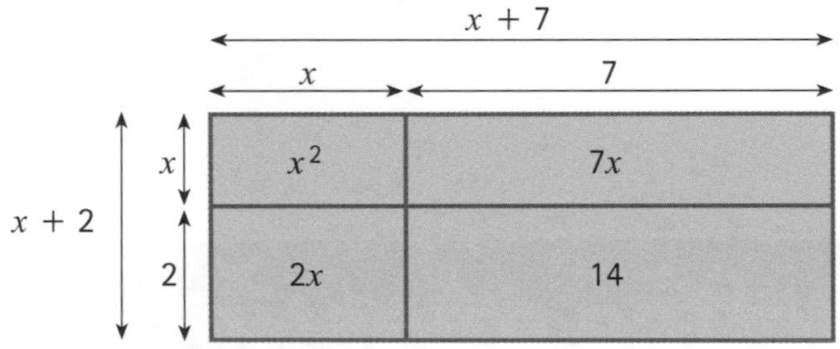

$(x + 7)(x + 2)$

$= x(x + 2) + 7(x + 2)$

$= x^2 + 2x + 7x + 14$

$= x^2 + 9x + 14$

	x	7
x	x^2	$7x$
2	$2x$	14

(Total width $x + 7$; total height $x + 2$.)

2. Expand and simplify:

$(x + 5)(x - 2) = x(x - 2) + 5(x - 2)$

Here we are taking every term in the first bracket and multiplying the second bracket by it, in turn.

$= x^2 - 2x + 5x - 10$

$= x^2 + 3x - 10$

3. Expand and simplify:

$(x + 7)(3 - x) = x(3 - x) + 7(3 - x)$

$= 3x - x^2 + 21 - 7x$

$= 21 - 4x - x^2$

C Expanding a squared expression

1 Expand and simplify:

$$(x+4)^2 = (x+4)(x+4)$$
$$= x(x+4) + 4(x+4)$$
$$= x^2 + 4x + 4x + 16$$
$$= x^2 + 8x + 16$$

2 Expand and simplify:

$$(x-9)^2 = x(x-9) - 9(x-9)$$
$$= x^2 - 9x - 9x + 81$$
$$= x^2 - 18x + 81$$

>> **key fact**

In general $(x+a)^2 = x^2 + 2ax + a^2$ and
$(x-a)^2 = x^2 - 2ax + a^2$

D The difference of two squares

1 Expand and simplify:

$$(x+9)(x-9) = x(x-9) + 9(x-9)$$
$$= x^2 - 9x + 9x - 81$$
$$= x^2 - 81$$

2 This is a special expansion that you must learn.
It is called the **difference of two squares**.

>> **key fact**

In general $(x+a)(x-a) = x^2 - a^2$

>> practice questions

Expand the following expressions.

1	$(x+3)(x+1)$	**2**	$(x+7)(x+2)$	**3**	$(x+2)^2$
4	$(x+m)(x+n)$	**5**	$(x+4)(x+2)$	**6**	$(x-a)^2$
7	$(3-x)^2$	**8**	$(a-x)^2$	**9**	$(x-y)^2$
10	$(x+2)(x-2)$	**11**	$(x+3y)(x-3y)$	**12**	$(6-x)(x+6)$

Inequalities

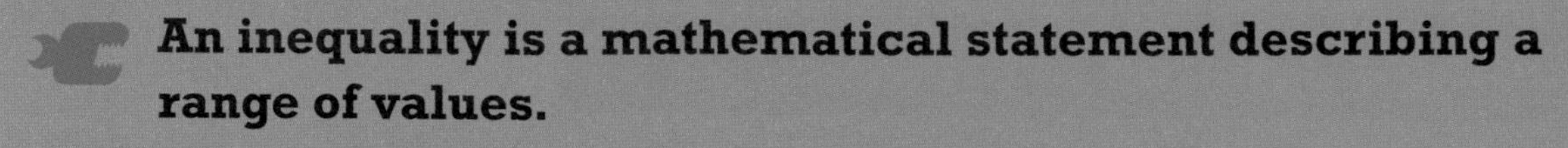

- An inequality is a mathematical statement describing a range of values.

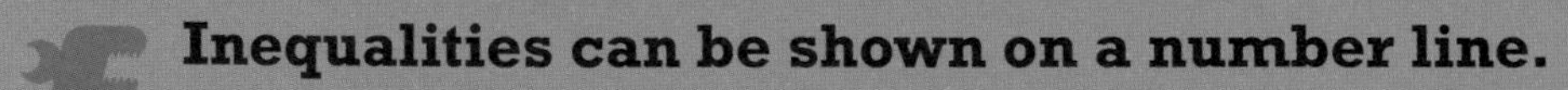

- Inequalities can be shown on a number line.

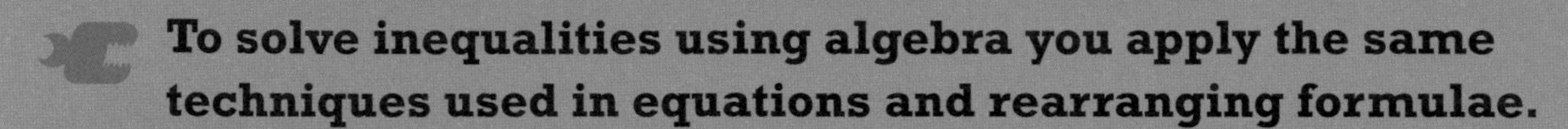

- To solve inequalities using algebra you apply the same techniques used in equations and rearranging formulae.

A Showing inequalities on a number line

There are four inequality symbols:

$>$ greater than	$\geqslant$ greater than or equal to
$<$ less than	$\leqslant$ less than or equal to

>> **key fact** **You can use a number line to show an inequality.**

1. We can show $x > 4$, which means x is greater than 4:

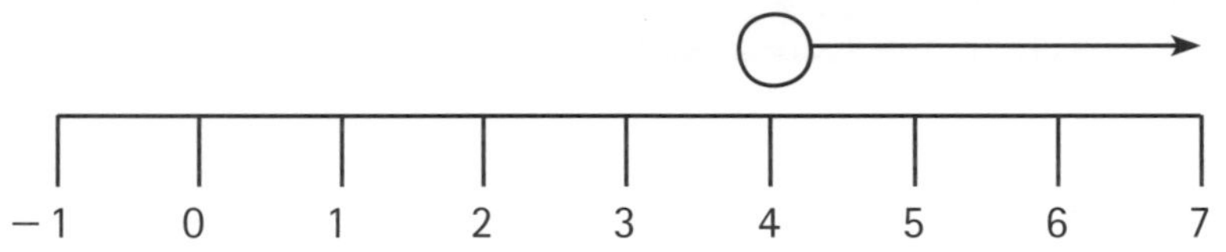

>> **key fact** **The empty circle means the value of 4 is not included in the inequality.**

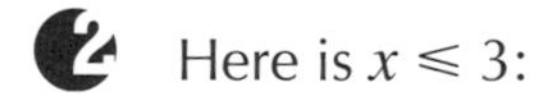

2. Here is $x \leqslant 3$:

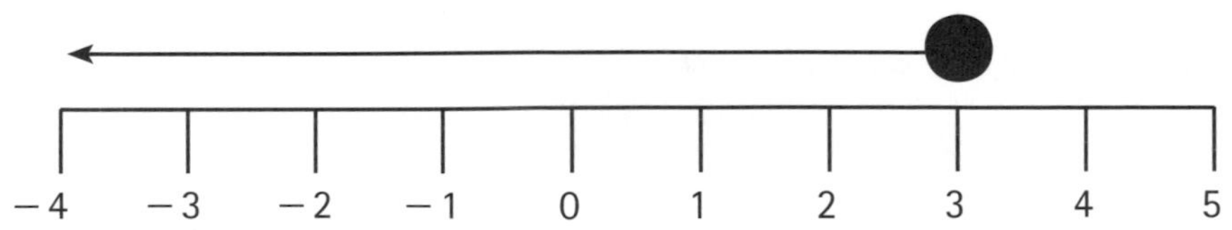

>> **key fact** **Here the filled circle means that $x = 3$ is included.**

3. Look at this line:

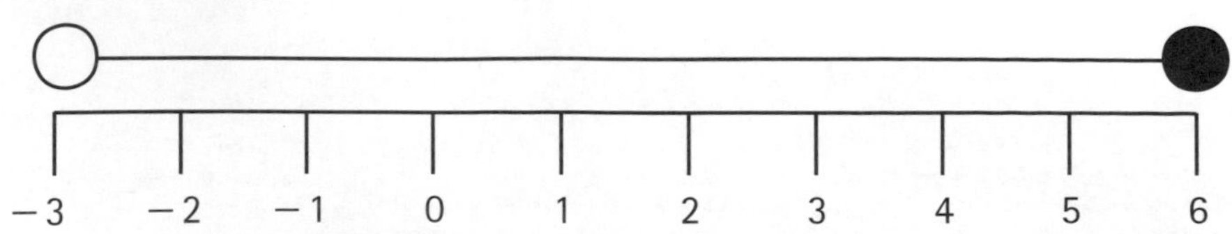

This line shows $-3 < x \leqslant 6$, this means x is between -3 and 6: it could be equal to 6 but cannot be equal to -3.

B Solving inequalities

>> **key fact** **You can solve equations to find an unknown number.**

>> **key fact** **You can solve inequalities to find a range of numbers.**

1. So $4 > -5$ is an inequality and it is a true statement.

But what happens if you:

(a) add 5 to both sides? (b) subtract 3 from both sides?

(c) multiply both sides by 6? (d) divide both sides by 2?

(e) multiply both sides by -4 (f) divide both sides by -2?

Investigate each of these statements and convince yourself.

You should find that when you multiply or divide by a negative quality, the direction of the inequality is **reversed**.

2. To solve this inequality:

$4x + 5 > 29$ *subtract 5 from both sides*

$4x + 5 - 5 > 29 - 5$

$4x > 24$ *divide both sides by 4*

$x > 6$

3. This is a bit more complex.

$-3 \leqslant 4x + 5 < 12$

Treat this as two inequalities.

Step 1	Work out $-3 \leqslant 4x + 5$	the answer is $-2 \leqslant x$
Step 2	Work out $4x + 5 < 12$	the answer is $x < \frac{7}{4}$
Step 3	Put them together	the final answer is $-2 \leqslant x < \frac{7}{4}$

>> practice questions

Solve these inequalities.

1 $2x + 7 \geqslant 3x + 2$

2 $7x + 3 > 5x - 4$

3 $8 - 6a \leqslant 7$

4 $1 - 6t \leqslant 9$

5 $-6d < 30$

6 $-2w \geqslant 5$

7 $\frac{1}{4}f + 3 \geqslant 1$

8 $6 > 2p + 3 > 4$

9 $1 \leqslant 5r + 2 \leqslant 12$

10 $-3 < \frac{1}{3}t + 2 < 5$

Number patterns and sequences

- **A sequence is a pattern of numbers that grows according to a mathematical rule.**
- **You can generate the terms of a sequence using a term-to-term rule or a position-to-term rule.**

A Standard number patterns

1. A sequence is a pattern of numbers that grows according to a mathematical rule. Each number in a sequence is called a **term**.
2. There are several patterns of numbers you need to be able to recognise when you see them. The '…' symbol means that the sequence carries on forever.

Name	Terms	Dot pattern
Even numbers	2, 4, 6, 8, 10, …	
Odd numbers	3, 5, 7, 9, 11, …	
Multiples of 3	3, 6, 9, 12, 15, …	
Square numbers	1, 4, 9, 16, 25, …	
Cube numbers	1, 8, 27, 64, 125, …	
Triangular numbers	1, 3, 6, 10, 15, …	
Powers of 2	2, 4, 8, 16, 32, 64, …	
Powers of 10	10, 100, 1000, 10 000, 100 000, 1 000 000, …	
Fibonacci numbers	0, 1, 1, 2, 3, 5, 8, 13, 21, …	

B Sequences from diagrams

1. You might be asked to produce a sequence of numbers from patterns in a diagram. You might also be asked to draw the next diagram in the sequence or predict the next number.
2. These diagrams generate the following sequences:

   White squares: 2, 4, 6, 8, 10, …

   Pink squares: 2, 6, 10, 14, 18, …

   Totals: 4, 10, 16, 22, 28, …

C Term-to-term rules

1 A **term-to-term** rule tells you how to find the next number in a sequence.

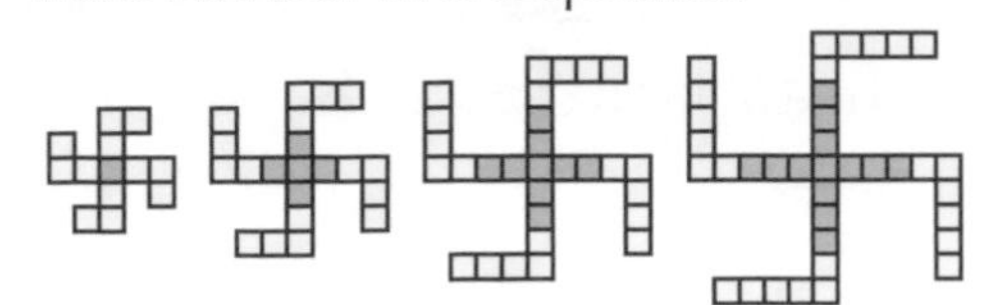

These diagrams generate the following sequences:

Green squares: 1, 5, 9, 13, ...
Yellow squares: 12, 16, 20, 24, ...

In both cases, the term-to-term rule is 'add 4'. The sequences are different because their first terms are different.

2 For the sequence of powers of 2, the term-to-term rule is 'multiply by 2'. However, starting with a number other than 2 produces a new sequence: 3, 6, 12, 24, 48, ...

3 More complex rules are possible. The rule 'divide by 2, then add 1', starting at 10, produces a sequence of decimals:
10, 6, 4, 3, 2.5, 2.25, 2.125, ...

4 Fibonacci sequences use the rule 'add the last two terms to find the next term'. You can start with any two numbers (e.g. 4, 7, 11, 18, 29, ...), but the most common one is given in section A.

D Position-to-term rules

1 You can generate the terms of a sequence from their **positions** in the sequence. The letter n is normally used for the position, and u_n for the nth term or general term.

>> **key fact** **The small number to the right of u is a suffix. It is just a sort of label and has nothing to do with indices.**

The following rule generates odd numbers.

Position	Calculation	Term
1	$2 \times 1 - 1$	$u_1 = 1$
2	$2 \times 2 - 1$	$u_2 = 3$
3	$2 \times 3 - 1$	$u_3 = 5$
n	$2 \times n - 1$	$u_n = 2n - 1$

The final cell in the table shows the position-to-term rule written in algebra. By substituting different numbers for n, you can find the value of any term in the sequence.
For example, for the 100th term, substitute $n = 100$: $u_{100} = 2 \times 100 - 1 = 199$.

2 The position-to-term rules for some of the sequences in section A are given below.

Even numbers	$u_n = 2n$
Multiples of 3	$u_n = 3n$
Square numbers	$u_n = n^2$
Cube numbers	$u_n = n^3$
Triangular numbers	$u_n = \frac{n(n+1)}{2}$
Powers of 2	$u_n = 2^n$
Powers of 10	$u_n = 10^n$

Check by substituting that they give the right results.

>> practice questions

1 Find the missing term(s) in each sequence.
(a) 4, 7, 10, ■, 16, ... (b) 10, 6, 2, ■, ...
(c) 5, 6, 8, 11, ■, ... (d) 83, ■, 74, 65, 53, ...
(e) ■, ■, 20, 40, 80, ... (f) 5, 9, 14, 23, ■, ...

2 Write down the first five terms of these sequences.

	(a)	(b)	(c)	(d)
First term	4	5	50	3
Term-to-term rule	add 10	subtract 4	divide by 2	multiply by 3 then subtract 4

3 Write down the first four terms of these sequences.
(a) $u_n = 4n - 2$ (b) $u_n = n^2 + 10$
(c) $u_n = 3n$ (d) $u_n = \frac{n(n+1)(n+2)}{3}$

Sequences and formulae

- **Use a difference table to find the rules for a sequence.**
- **Some sequences are created from standard number patterns by a simple operation.**
- **To find the formula for a sequence of fractions, analyse the numerators and denominators separately.**

A Difference tables

1 A **difference table** can make it easier to see how the numbers in a sequence are produced.

>> **key fact** **A difference table should show the positions, terms and differences.**

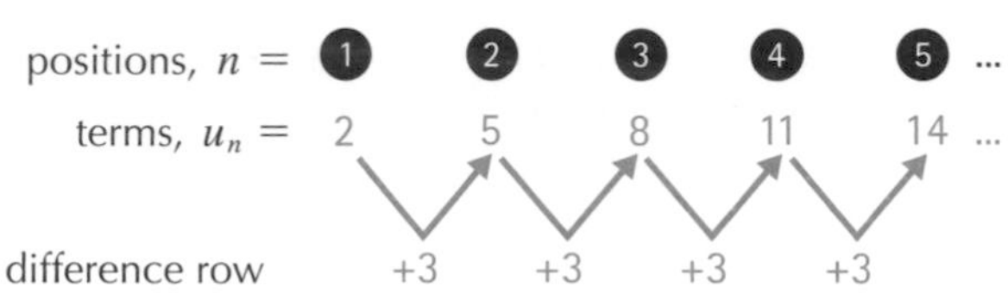

This shows that the terms increase by 3 each time.

B Linear sequences

1 **key fact** **If the numbers in the difference row are all the same, you are dealing with a linear sequence. This means that if you plotted the terms against their positions on a graph, all the points would be on a straight line.**

With a linear sequence, you can write down the term-to-term rule immediately. The rule for the sequence in section A is 'first term 2, add 3'.

2 The position-to-term rule must contain $3n$ in the formula to make all the differences 3. However, although this makes the terms increase in the right way, it doesn't give the first term as 2.

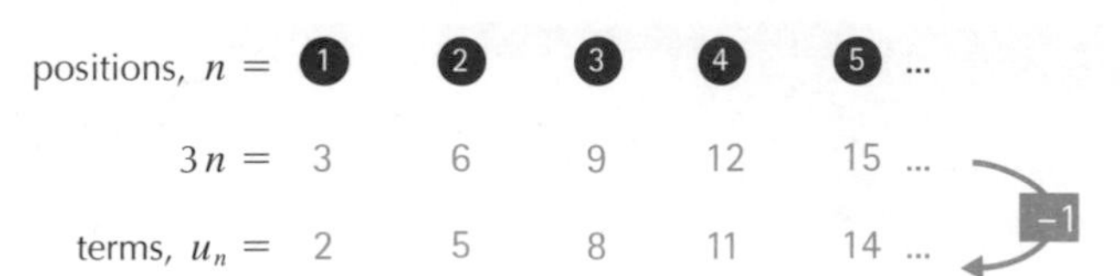

The numbers are all 1 less than a multiple of 3, so the formula is $u_n = 3n - 1$.

3

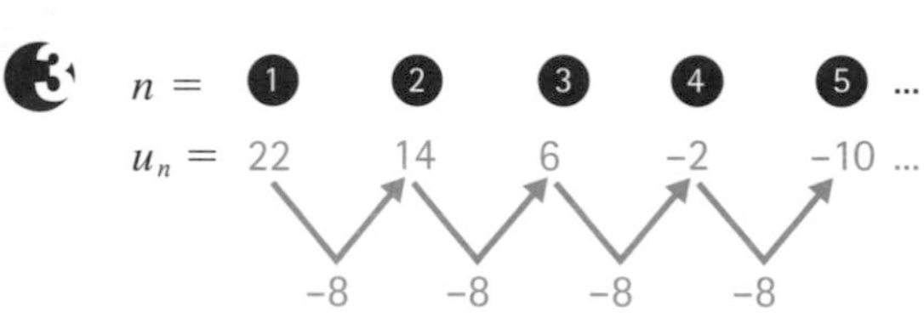

This shows that the terms decrease by 8 each time.

Here the differences are negative because the terms decrease. The term-to-term rule for this sequence is 'first term 22, subtract 8'.
The position-to-term rule needs to have $-8n$ in it. The formula for this sequence is $u_n = -8n + 30$

C Other sequence types

1 Sometimes, you may need a second difference row to see what is happening.

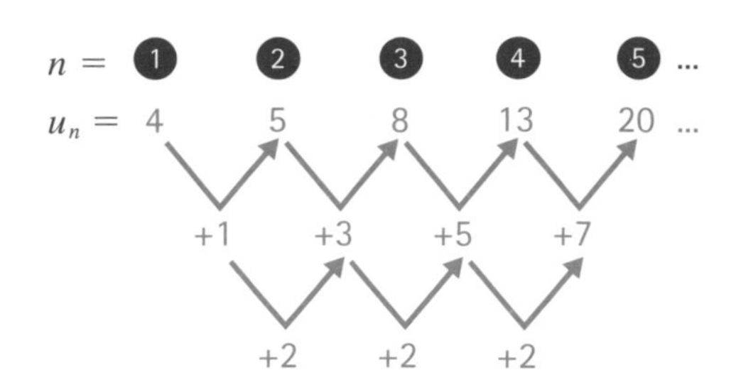

You could describe this as 'first term 4, add 1, 3, 5, 7, etc.' Sequences like this do have position-to-term formulae, but you are not required to find them in the Foundation tier exam.

If the differences don't reveal a simple pattern, check these two possibilities.

2

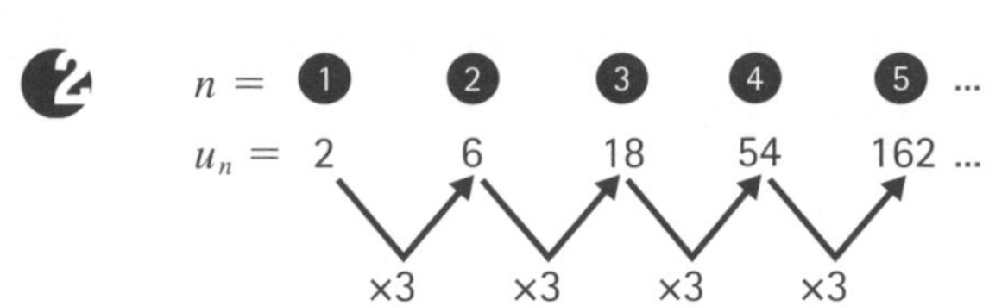

This is a sequence involving powers, as you multiply by the same number each time. The term-to-term rule is 'first term 2, multiply by 3'. The position-to-term formula is $u_n = \frac{2\times3^n}{3}$, or $u_n = 2 \times 3^{(n-1)}$.

3

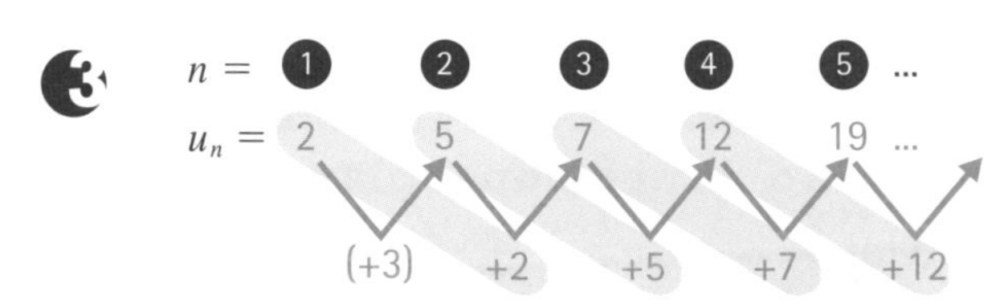

If the differences turn out to be the same as the terms, you are dealing with a Fibonacci-type sequence. The rule is 'add the previous two terms', and you have to state the first **two** terms. This can also be written $u_n = u_{n-1} + u_{n-2}$.

D Variations on standard patterns

1 Some sequences can be built up by altering the standard number patterns slightly.

Examples:

2, 5, 10, 17, 26, ...	square numbers with 1 added	$u_n = n^2 + 1$
2, 8, 18, 32, 50, ...	square numbers doubled	$u_n = 2n^2$
11, 18, 37, 74, 135, ...	cube numbers with 10 added	$u_n = n^3 + 10$
20, 40, 80, 160, 320, ...	powers of 2 multiplied by 10	$u_n = 10 \times 2^n$

E Sequences of fractions

1 Analyse the numerator and denominator separately, then combine them into a single formula.

Example: $\frac{2}{1}, \frac{5}{4}, \frac{8}{9}, \frac{11}{16}, \frac{14}{25}, \ldots$

The numerators are the terms of the sequence in section A, $u_n = 3n - 1$.

The denominators are square numbers, $u_n = n^2$.

So the fraction sequence is $u_n = \frac{3n-1}{n^2}$.

>> practice questions

Each question gives the first five terms of a sequence. For each one:

(a) Draw a difference table and find a term-to-term rule (if possible).

(b) Find a position-to-term formula (if possible).

(c) Calculate the 10th term.

1 6, 11, 16, 21, 26, ...

2 3, 9, 15, 21, 27, ...

3 10, 9, 8, 7, 6, ...

4 1, 1.2, 1.4, 1.6, 1.8, ...

5 2, 5, 9, 14, 20, ...

6 12, 48, 192, 768, 3072, ...

7 0, 3, 8, 15, 24, ...

8 2, 16, 54, 128, 250, ...

9 11, 17, 28, 45, 73, ...

10 $\frac{1}{2}, \frac{3}{4}, \frac{5}{8}, \frac{7}{16}, \frac{9}{32}, \ldots$

Co-ordinates

- **Co-ordinates describe positions on a two-dimensional grid.**
- **Positive co-ordinates are used for positions above or to the right of the origin, negative ones below or to the left.**
- **The axes divide the grid into four quadrants.**

A Working on a grid

1. **key fact** **Co-ordinates are used to describe positions on a flat, two-dimensional plane.**

Two lines with scales, the **axes**, create a **grid** of squares. They cross at a point called the **origin**. The ***x*-axis** describes distances to the right of the origin, and the ***y*-axis** distances above the origin.

2. To reach the point P from the origin, you have to travel 2 grid units to the right and then 5 up. This is written (2, 5). To reach Q, you have to travel 4 units to the right and then 1 up, so Q is the point (4, 1). You say that the x co-ordinate of Q is 4 and its y co-ordinate is 1. The origin's co-ordinates are (0, 0).

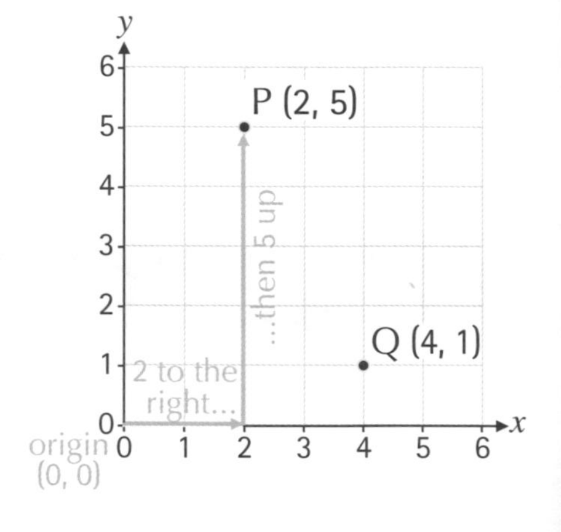

3. To remember the order (right first, then up), you could memorise the phrase 'along the corridor before you climb the stairs'.

>> **key fact** **Points that are on an axis have a zero in their co-ordinates.**

B The four quadrants

1. Co-ordinates wouldn't be much use if you could only talk about points to the right and above the origin.

>> **key fact** **Negative numbers are used to describe distances to the left and below the origin.**

2.

T (−3, 4)
U (2,−2)
V (−5,−4)

3. When you include positive and negative co-ordinates, you divide the grid into four **quadrants**.

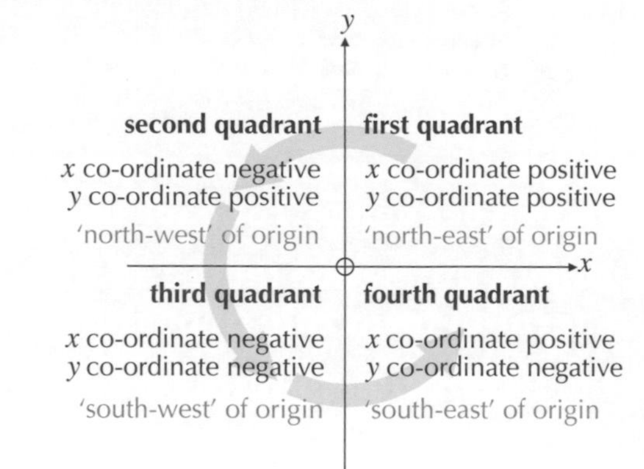

C Co-ordinate problems

1. Questions involving co-ordinates often ask you to complete a shape.

2. Here you have to add the fourth vertex to make a rectangle.

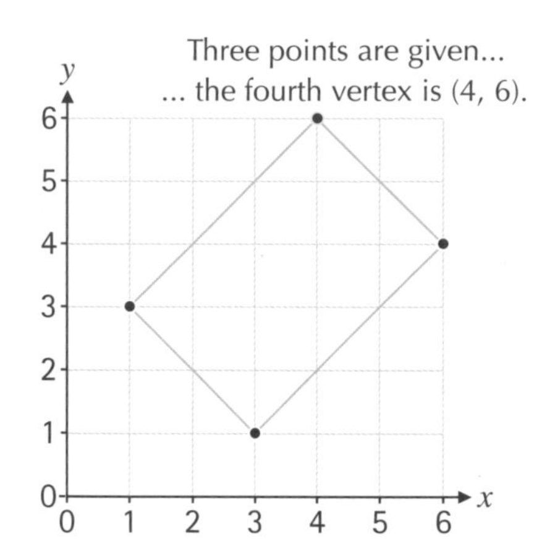

3. In this question, you have to add another point to make an isosceles triangle.

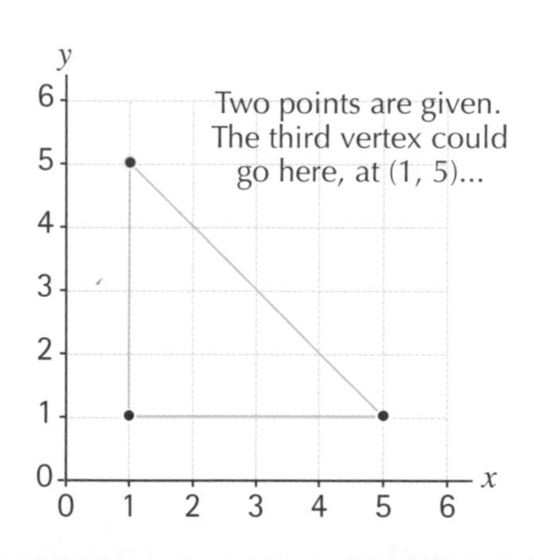

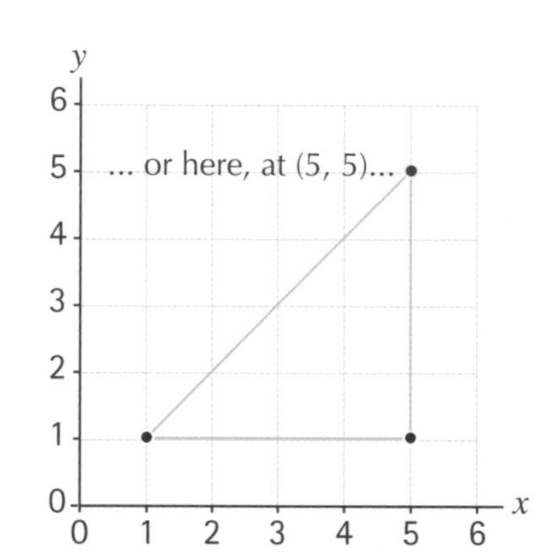

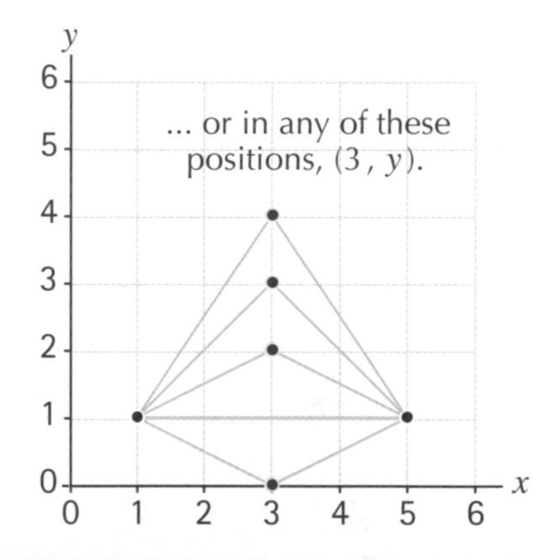

>> practice questions

1 Write down the co-ordinates of each point marked on the grid.

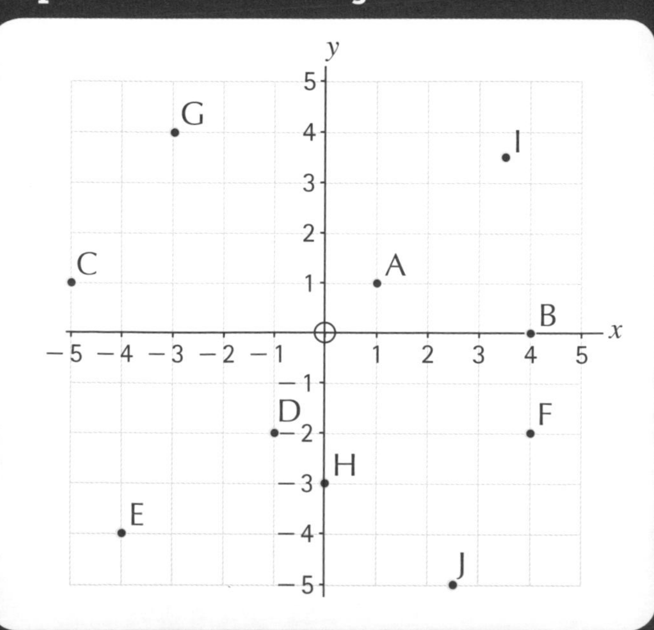

2 In section C3 above, only the first quadrant was shown. There are more positions for the third vertex if the fourth quadrant is shown as well. Re-draw the grid and find at least four extra positions.

3 Draw a co-ordinate grid with axes going from −10 to 10 in both directions.

Mark the point M at (2, 1). M is one vertex of a rectangle whose area is 6 square units. Find all the possible positions for the vertex opposite to M (hint: there are sixteen!).

Lines and equations

$y = c$ is the equation of a horizontal straight line.
The equation of the x-axis is $y = 0$.

$x = c$ is the equation of a vertical straight line.
The equation of the y-axis is $x = 0$.

All other straight lines have equations of the form $y = mx + c$, where m is the gradient of the line and c is the y-intercept.

A Gradients and y-intercepts

1 Look at this graph.
It shows the equations:

$y = 2x + 4$ $\quad$ $y = 2x + 2$

$y = 2x + 3$ $\quad$ $y = 2x + 1$

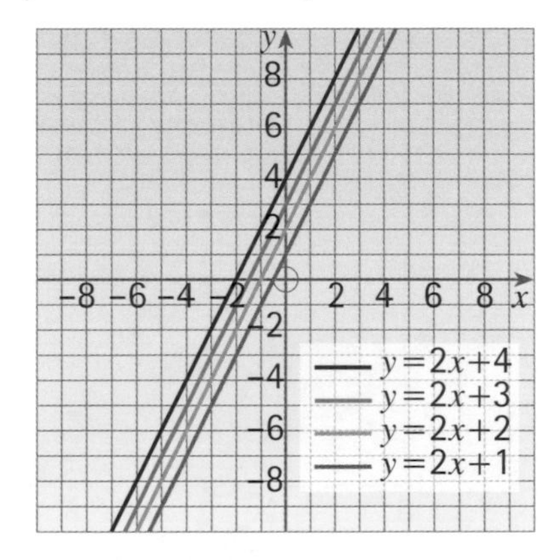

All four lines are parallel. Notice that the number in front of the x is 2 in each case (this is the **coefficient** of x). This is the **gradient** of the line. The second number in the equation gives the y-coordinate of the point where the line cuts the y-axis. This point is called the **y-intercept**.

>> **key fact** **Find the y-intercept by substituting $x = 0$ into the equation.**

2 *Example*: In the equation, $y = 3x + 2$, the y-axis is where $x = 0$.

So the y-intercept is at:
$y = (3 \times 0) + 2$ or more simply $(0, 2)$.

B Some examples

 $y = 9x - 5$

Write down:

(a) the gradient – this is the coefficient of x (the number in front of the x and consequently this must be 9).

(b) the y-intercept of the line – this is where the line cuts the y-axis. The y-axis is the line where $x = 0$. By substituting $x = 0$ into the equation, we can calculate the value of y. This means that y is -5.

So the y-intercept is at $(0, -5)$.

 $y = -5x + 7$

Write down:

(a) the gradient – again the gradient is the coefficient of x, so it is -5.

(b) the y-intercept of the line – put zero into the equation for x and the y-intercept is $(0, 7)$.

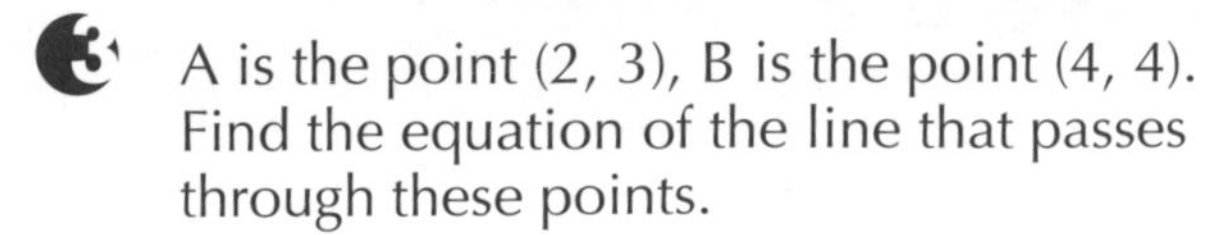

3 A is the point (2, 3), B is the point (4, 4). Find the equation of the line that passes through these points.

The gradient, m, is the measure of the slope. To work this out use the equation:

$$m = \frac{\text{increase in } y \text{ values}}{\text{increase in } x \text{ values}}$$

$$m = \frac{4-3}{4-2}$$

$$m = \frac{1}{2}$$

To find c we substitute what we know into

$$y = mx + c$$

$$y = \frac{1}{2}x + c$$

We have the values of x and y from the points the line passes through.

Using (2, 3) we can say $3 = \frac{1}{2} \times 2 + c$.

Solving this equation gives us $c = 2$.

So our required equation is $y = \frac{1}{2}x + 2$.

C Drawing graphs of equations

1 Draw the line with the equation $y = 2x + 1$.

2 Work out three points

when $x = 1$, $y = (2 \times 1) + 1 = 2 + 1 = 3$

when $x = 2$, $y = (2 \times 2) + 1 = 4 + 1 = 5$

when $x = 3$, $y = 7$.

So the 3 points are (1, 3), (2, 5), (3, 7).

3 Write down the equation of the line parallel to $y = 2x + 1$, whose intercept is $(0, -3)$.

The line parallel to $y = 2x + 1$, with a y-intercept (0, 13) is a line with the same gradient, so it must have a 2 in front of the x and it must cut the y-axis at -3. This means that the equation of the line must be $y = 2x - 3$.

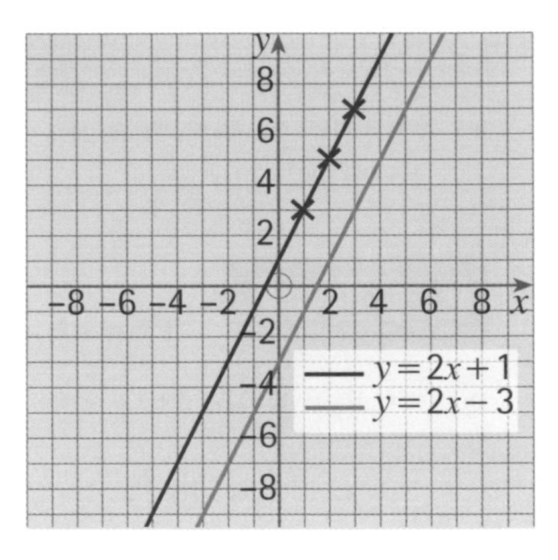

>> practice questions

Write the gradient and the y-intercept of lines with the following equations:

1 $y = 7x + 3$ **2** $y = 3x - 5$ **3** $y = \frac{1}{3}x + 12$

4 $y = 3x - 2$ **5** $3x + y = 4$ **6** $2x - y = 7$

7 **The gradient of a line is 4 and its y-intercept is (0, 5). What is the equation of the line?**

8 **A line passes through (0, 6) and (2, 8). Find the equation of the line.**

9 **Find the equation of the line that passes through the point (0, 1) and is parallel to $y = 2x - 1$.**

Quadratic graphs

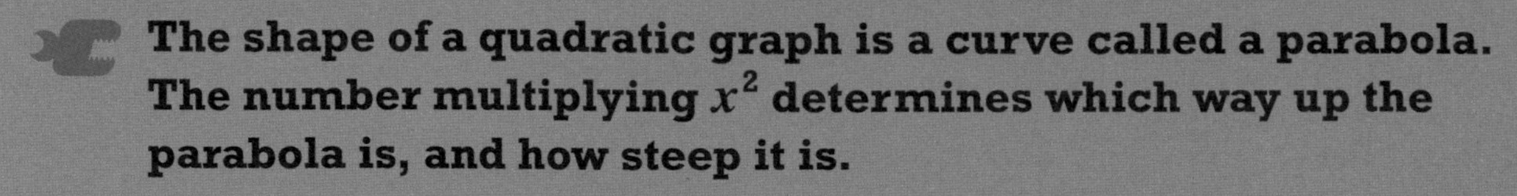

- **The shape of a quadratic graph is a curve called a parabola. The number multiplying x^2 determines which way up the parabola is, and how steep it is.**
- **A quadratic equation can be solved by finding where its graph crosses the x-axis.**

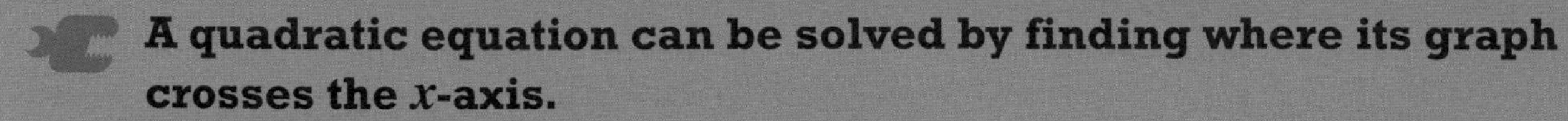

A The graph of $y = x^2$

1 To plot this graph, follow this plan:

- Make a table for x and y values.
- Choose x values for the range you need (e.g. -3 to 3).
- Square the x values to calculate the y values.

x	-3	-2	-1	0	1	2	3
$y = x^2$	9	4	1	0	1	4	9

- Plot the points. Join them with a smooth curve.

2 **key fact** **The shape of the graph of $y = x^2$ is called a parabola.**

Notice that it has the y-axis as a line of symmetry, and passes through the origin.

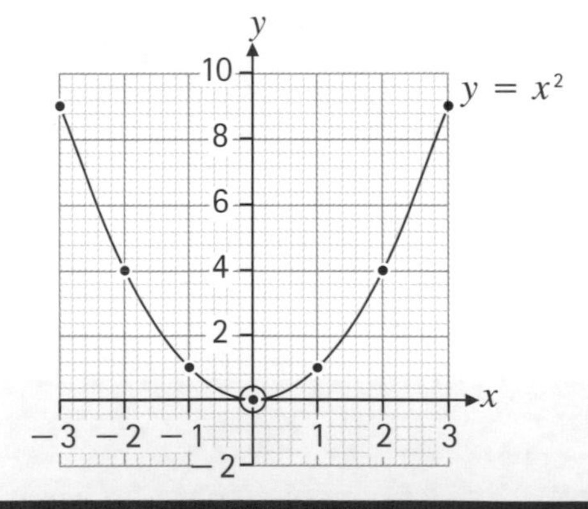

3 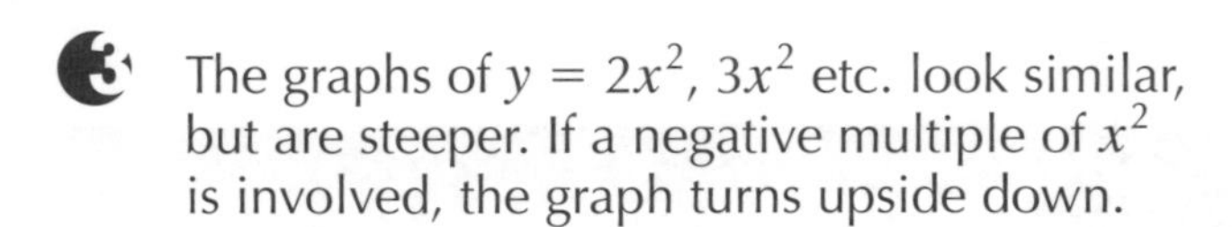

The graphs of $y = 2x^2$, $3x^2$ etc. look similar, but are steeper. If a negative multiple of x^2 is involved, the graph turns upside down.

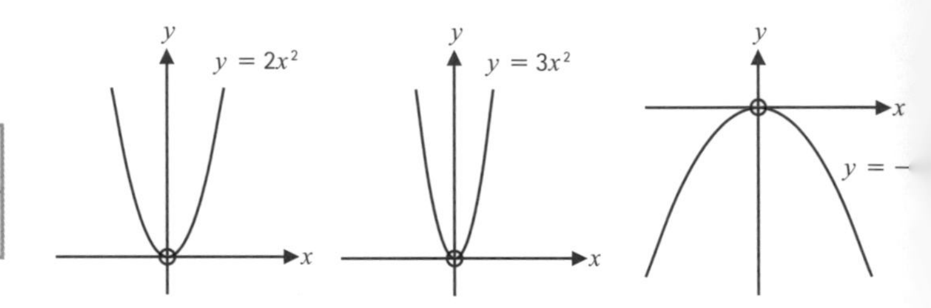

B Other quadratic graphs

1 The equation of a quadratic graph may also contain multiples of x and numbers. These graphs don't have to pass through the origin and may have a different line of symmetry.

2 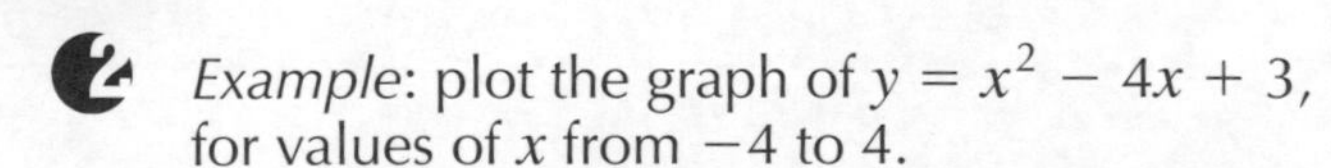

Example: plot the graph of $y = x^2 - 4x + 3$, for values of x from -4 to 4.

Prepare a table and calculate the y values. Plot the points and join them with a smooth curve.

x	-4	-3	-2	-1	0	1	2	3	4
$y = x^2 - 4x + 3$	35	24	15	8	3	0	-1	0	3

The line of symmetry of this graph is $x = 2$.

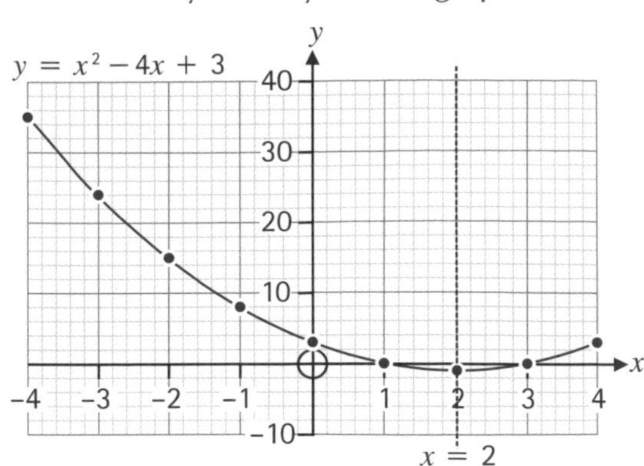

 You may be given the equation in factorised form, such as $y = x(x - 2)$ or $y = (x + 1)(x + 5)$. You can calculate the y values directly from these, or expand them first if you prefer.

C Solving quadratic equations

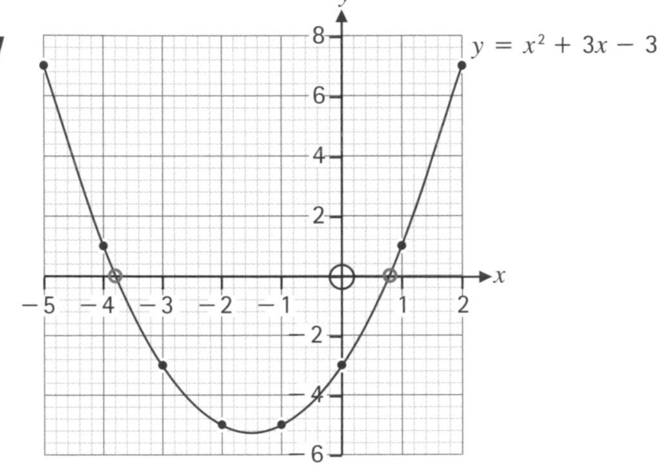

1 A quadratic equation contains an expression with x^2, x and numbers, set equal to zero. You can solve this kind of equation using a graph.

Example:

Solve $x^2 + 3x - 3 = 0$, plotting for x values between -5 and 2. Give your answers correct to 1 decimal place. You need to draw the graph of $y = x^2 + 3x - 3$. Here is the table of values.

x	-5	-4	-3	-2	-1	0	1	2
$y = x^2 + 3x - 3$	7	1	-3	-5	-5	-3	1	7

Plot the points.

The solutions to the equation are found where the graph crosses the x-axis. You can read the solutions off directly. The solutions to this equation are $x = 0.8$ and $x = -3.8$, to 1 dp.

 Quadratic equations may have different numbers of solutions.

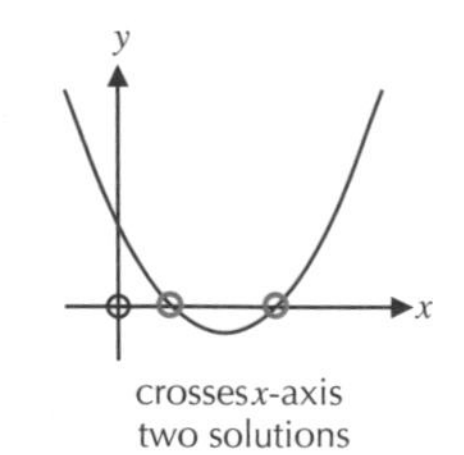

crosses x-axis two solutions

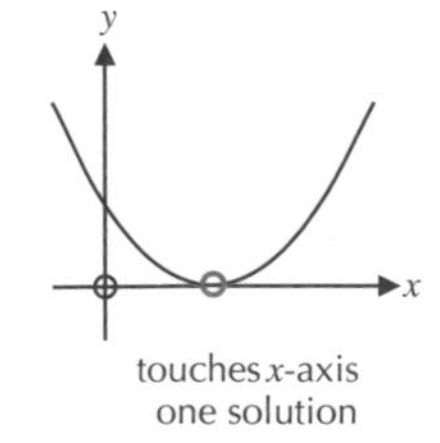

touches x-axis one solution

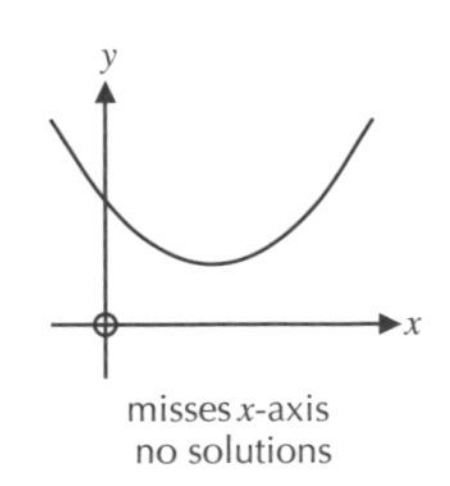

misses x-axis no solutions

>> practice questions

1 Plot the following quadratic graphs. Use x values from -5 to 5.

(a) $y = x^2 + 5$ **(y-axis: 0 to 30)** **(b)** $y = 2 + 3x - x^2$ **(y-axis: -40 to 10)**

(c) $y = \frac{1}{2}x^2 + x$ **(y-axis: 0 to 30)** **(d)** $y = (x - 2)(2x + 1)$ **(y-axis: -10 to 70)**

2 Solve these equations graphically, giving your answers correct to 1 dp.

(a) $x^2 - 2x - 1 = 0$ **(x-axis: -3 to 5; y-axis: -5 to 15)**

(b) $3x^2 - 10 = 0$ **(x-axis: -3 to 3; y-axis: -10 to 20)**

Measures and accuracy

Metric units can take a range of prefixes. These change the size of the unit by a power of ten.

Imperial units are not related to each other in tens, but many are still in use.

A measurement accurate to the nearest unit may be out by up to half a unit above or below.

A Units of measure

1. We use standard units of measurement so that everyone agrees on the size of an object, an amount of liquid, and so on.
2. In the metric system, the most common units are shown in the table.
3. Capacity is actually a measure of volume, usually applied to quantities of liquid.

Quantity	Unit(s)
length	metre (m)
mass	gram (g), tonne (t)
area	square metre (m^2), acre (a)
volume	cubic metre (m^3)
capacity	litre (l)
electrical quantities	Volt (V), Ampère (A), Watt (W)

B Metric prefixes

1. Each metric unit can be altered in size by adding a prefix. This table illustrates the most usual prefixes applied to the metre.

prefix	length unit	abbreviation	meaning
milli-	millimetre	mm	$\frac{1}{1000}$ of a metre
centi-	centimetre	cm	$\frac{1}{100}$ of a metre
kilo-	kilometre	km	1000 metres

So 10 mm = 1 cm,
100 cm = 1 m and
1000 mm = 1 m.

2. There are some other relationships you need to know:

- 1000 kilograms (kg) = 1 tonne
- 1 cubic centimetre (cm^3) = 1 millilitre (ml)
- 1000 cm^3 = 1 litre and 1000 l = 1 cubic metre (m^3)
- 10 000 square metres = 1 hectare (used for areas of land).

C Imperial units

1. The older imperial units are still in use, particularly for distances on road signs.

Length

imperial	metric
1 inch (in or ″)	2.5 cm
1 foot (ft or ′) = 12 in	30 cm
1 yard (yd) = 3 ft	90 cm
1 mile (mi) = 1760 yd	1.6 km

Mass

imperial	metric
1 ounce (oz)	28.4 g
1 pound (lb) = 16 oz	454 g
1 stone (st) = 14 lb	6.36 kg
1 ton = 2240 lb	≈1 tonne

Capacity

imperial	metric
1 pint	57 cl
1 gallon = 8 pints	4.6 l

D Accuracy

1. Suppose that you are told that a length is 63 mm, correct to the nearest millimetre. There is a range of actual measurements that could have produced this figure. Anything 63.5 mm or above would round up to 64 mm. Anything below 62.5 mm would round down to 62 mm. So the range is 62.5 mm ≤ length < 63.5 mm. Notice that 62.5 mm is allowed, but 63.5 is not.

>> **key fact** **If a measurement is given to the nearest unit, the tolerance is half a unit each way.**

2. Here are some further examples showing different levels of accuracy. The symbol ± means 'plus or minus'.

length	accuracy	tolerance	range allowed
450 mm	nearest 10 mm	± 5 mm	445 mm ≤ length < 455 mm
98 m	nearest metre	± 0.5 m	97.5 m ≤ length < 98.5 m
6.33 km	2 decimal places	± 0.005 km	6.325 km ≤ length < 6.335 km
210 cm	nearest 5 cm	± 2.5 cm	207.5 cm ≤ length < 212.5 cm

3. If you use rounded measurements in a calculation, a range of answers is possible. Suppose you had a rectangle 8 cm by 6 cm, but the measurements were known to be accurate to only the nearest centimetre. What range of area could the rectangle have?

7.5 cm ≤ length < 8.5 cm   5.5 cm ≤ length < 6.5 cm

The minimum possible area is $7.5 \times 5.5 = 41.25\ \text{cm}^2$.

The maximum possible area is $8.5 \times 6.5 = 55.25\ \text{cm}^2$.

So $41.25\ \text{cm}^2 \leq \text{area} < 55.25\ \text{cm}^2$.

>> practice questions

1 Rewrite each measurement in the unit given.

(a) 3 km ▶ m (b) 50 ml ▶ cl (c) 4.2 kg ▶ g

(d) 31.2 cl ▶ l (e) 7 t ▶ kg (f) 2 cm² ▶ mm²

(g) 2 pt ▶ l (h) 9 st ▶ kg (i) 300 g ▶ oz

(j) 5 in ▶ mm (k) 240 km ▶ mi (l) 10 kg ▶ lb

2 Write down the range of measurements for each rectangle, then calculate its smallest and largest possible area.

(a) 10 cm by 5 cm, accurate to the nearest cm

(b) 300 m by 250 m, accurate to the nearest 10 m

(c) 16 mm by 6.5 mm, accurate to the nearest 0.5 mm

Dimensions

All formulae for length or distance have one dimension only.

All formulae for area have 2 dimensions of length.

All formulae for volume have 3 dimensions of length.

A Length

You can check if a formula is describing length by looking at its components.

Consider a square of side length p. The formula for the perimeter is $4p$.

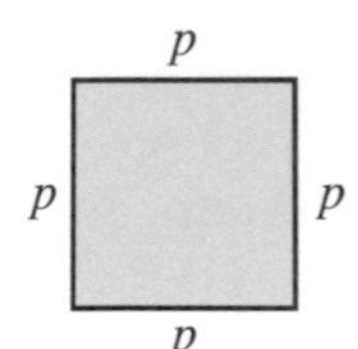

This formula obviously represents a length, because p is a length and 4 is just a number.

A rectangle's perimeter will also have a formula for the sum of the lengths of its sides.

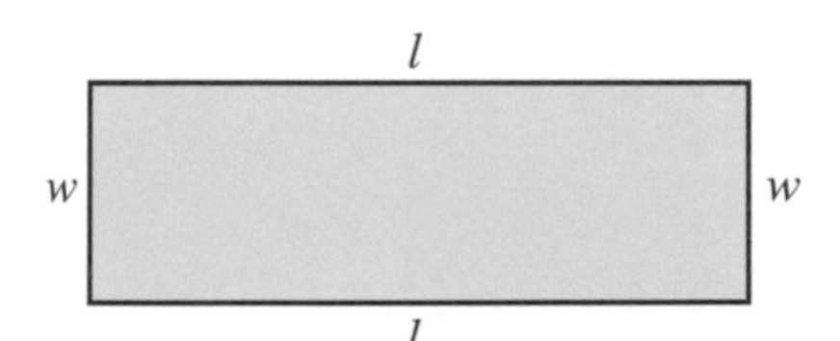

The formula here becomes $2l + 2w$ or $2(l + w)$.

Again, this is a number multiplied by a length.

3 A circle of diameter d units has a perimeter of πd or $2\pi r$.

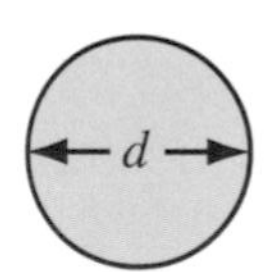

4 Rob says that the formula $3\pi r$ represents a length, but John argues that it cannot be so. Who is correct?

3 and π are numbers, they do not have a dimension, therefore r is the only element of the formula that does have a dimension. That means this formula is of **dimension one**, so it *may* be a length.

B Area

>> **key fact** **Whenever you have a formula for area, you will always have: length × length.**

1. The formula for the area of a circle is πr^2.

   In the analysis you can ignore π because it is a number and therefore has no dimension, but you do have $r \times r$ in this formula.

2. In other words, you have an area. So the formula is of **dimension two**.

C Volume

>> **key fact** **All formulae for volume have the dimensions length × length × length.**

1. The formula for the volume of a cylinder is $\pi r^2 h$.

   Again we can simply ignore the numbers, in this case π, so the dimensions to consider are r^2 and h.

   So this is written as $r \times r \times h$, so it must be a volume.

2. *Example*:

   $5\pi r^2 \qquad \frac{4}{3}\pi r^3 \qquad \frac{3}{5}\pi r^2$.

   Which one is a volume?

   First of all, ignore the numbers because they do not have dimensions.

   So in $5\pi r^2$ we need to consider $r \times r$. This is **dimension two**, so it must be an area.

   This is also true for $\frac{3}{5}\pi r^2$.

   In $\frac{4}{3}\pi r^3$, we need to consider $r \times r \times r$. This is **dimension three**, so this must be a volume.

>> practice questions

1 The letters h and r represent lengths. For each of the following formulae, write down if it represents a length, an area or a volume.

(a) $2\pi r$ **(b)** πr^2

(c) $\pi r^2 h$ **(d)** $2\pi rh$

2 The letters l and w represent lengths. Explain why $l^2 lw$ cannot represent an area.

3 Explain what the formula $\frac{4}{3}\pi r^3$ represents.

Time

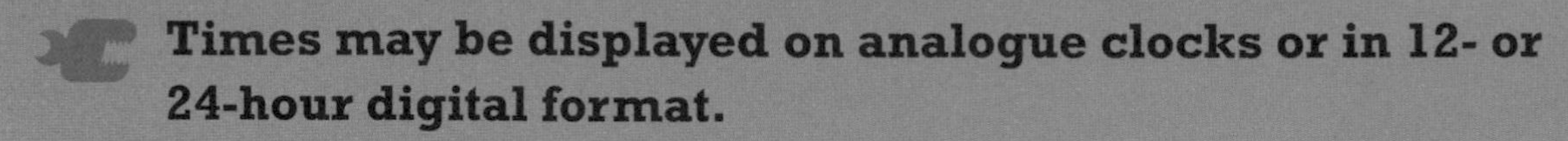
- **Times may be displayed on analogue clocks or in 12- or 24-hour digital format.**

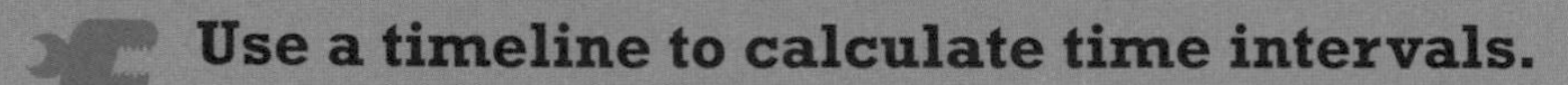
- **Use a timeline to calculate time intervals.**
- **Memorise simple fractions of an hour and a minute.**

A Units of time

1 **key fact** **Time units do not relate to each other in tens or hundreds.**

This makes them different from the metric units we use, and can make them difficult to work with.

1 day = 24 hours	… = 1440 minutes	… = 86 400 seconds
	1 hour = 60 minutes	… = 3600 seconds
		1 minute = 60 seconds

These units are often shortened to 'hour, min, sec' or even 'h, m, s'.

It is useful to memorise simple fractions of an hour or minute.

fraction of an hour	$\frac{1}{2}$	$\frac{1}{4}$	$\frac{3}{4}$	$\frac{1}{3}$	$\frac{2}{3}$	$\frac{1}{5}$	$\frac{1}{10}$
minutes	30	15	45	20	40	12	6

B Time displays

1 We are used to seeing time displayed in many different ways.

	morning	midday	afternoon	evening	night	midnight
12-hour digital	8.25 am	12 noon	3.40 pm	7 pm	10.30 pm	12 midnight
analogue						
24-hour digital	08:25	12:00	15:40	19:00	22:30	00:00

2 You may need to change a time from 12-hour to 24-hour. If the time is after noon and before midnight (a **pm** time), add 12 to the hours. So 2.15 pm = 14:15, but 2.15 am = 02:15.

To change back, do the opposite.

C Time intervals and timelines

1 **key fact** **A time interval is the difference between two given times.**

2 A **timeline** can help you to calculate a time interval. Suppose you wanted to know how long it was from 11.45 am until 2.20 pm. Write the times out on a line like this:

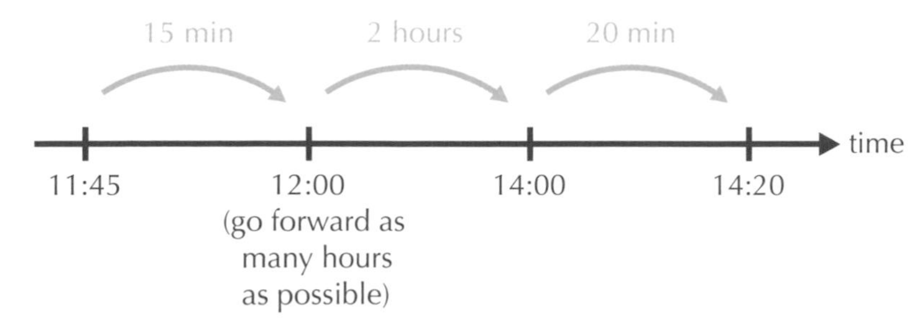

The time interval is the sum of all the parts: 15 min + 2 hours + 20 min = 2 hours 35 min.

D Time calculations

1 Suppose the time is 21:32 and you need to know what time it will be 5 hours 47 minutes later.

	h	*m*
	21	32
+	5	47
	26	79

79 minutes is 1 hour 19 min, so the answer is 27 hours 19 min. The hours have gone over 24, so subtract 24 to give 3. The time is 03:19.

2 You can multiply times in a similar way. Suppose an assembly line took 1 hour 25 minutes to assemble a machine. How long would it take to assemble six machines?

	h	*m*
	1	25
×		6
	6	150

150 minutes = 2 hours 30 min, so the answer is 8 hours 30 min.

3 Warning! You can't use your calculator to do a question like this. It's quite common for people to try to do it by entering 1.25 × 6, getting 7.5, then answering 7 hours 5 minutes!

4 Suppose 5 hours 20 minutes of teaching time has to be divided into 8 lessons. Rewrite the time in the smallest possible unit: 5 h 20 min = 320 min. 320 ÷ 8 = 40 min per lesson.

>> practice questions

1 Copy and complete the table.

12-hour	3 pm	2.25 pm			4.45 am		7.15 pm
24-hour	15:00		06:40	22:10		00:05	

2 Rewrite each time in the units given.

(a) 3 hours ▶ minutes (b) 12 min ▶ sec (c) $2\frac{1}{2}$ hours ▶ min ▶ sec (d) $4\frac{1}{4}$ hours ▶ min ▶ sec
(e) 500 sec ▶ min and sec (f) 448 min ▶ h and min (g) 2.1 hours ▶ min (h) 15.8 min ▶ sec

3 Find the time interval between the given times.

start	13:05	09:40	10.50 pm	16:22
end	16:25	11:15	2.05 am	18:18

4 Carry out the following calculations.

(a) What time is 3 h 25 min after 2.50 pm? (b) How long is 10 intervals of 32 seconds?

(c) Divide $6\frac{1}{2}$ hours into 5 equal length intervals.

Compound measures

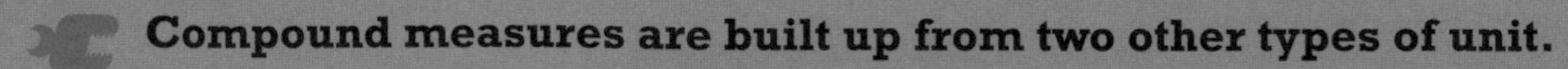

- **Compound measures are built up from two other types of unit.**
- **Average speed = $\frac{\text{distance}}{\text{time}}$; Density = $\frac{\text{mass}}{\text{volume}}$.**

A Speed

1 Speed is measured in a variety of different units: miles per hour (mph), kilometres per hour (km/h), metres per second (m/s), etc.

>> **key fact** **All units of speed are created from a length unit and a time unit.**

2 During a journey, the speed of a moving object changes as it accelerates and decelerates. Exam questions are usually about the **average** speed for a journey.

>> **key fact** ***Example*: A train takes 2 hours to cover a distance of 130 km. What is its average speed?**

Speed = $\frac{130}{2}$ = 65 km/h.

3 When working with speed, you have to be careful that the units match.

Example:

Ellie cycles 3 km to school in 15 minutes. What is her average speed in km/h?

15 minutes = $\frac{1}{4}$ hour (match the time units in the question and the answer)

Speed = $3 \div \frac{1}{4} = 3 \times 4 = 12$ km/h.

If you just went ahead and divided the numbers given in the question, you would get $3 \div 15 = 0.2$, but of course, this is in kilometres per minute!

4 You may need to convert one unit of speed to another.

Example:

1 metre per second = 60 metres per minute = 3600 metres per hour = 3.6 km/h.

B Time and distance

1 There are two other formulae you may need to use:

>> **key fact** **time = $\frac{\text{distance}}{\text{speed}}$**

distance = speed × time

2 You can remember this using the DST triangle:

To use it, cover up the letter for the quantity you're trying to find. The other two are in the right relationship, i.e. this one shows 'time = $\frac{\text{distance}}{\text{speed}}$'.

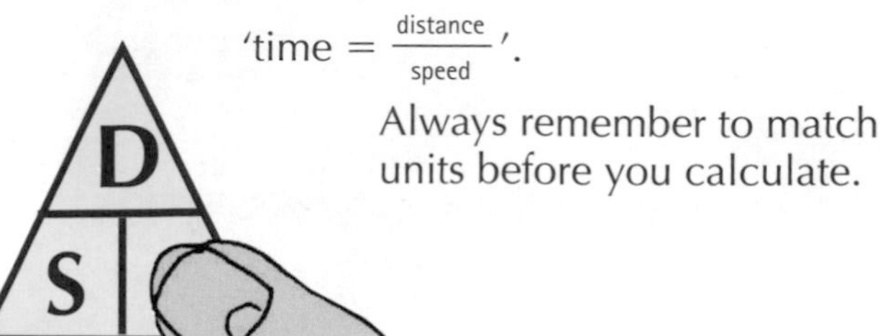

Always remember to match units before you calculate.

3 *Example*:

Adil drives for $3\frac{1}{2}$ hours at an average speed of 50 mph. How far was his journey?

Distance = $50 \times 3.5 = 175$ miles.

Example:

The average speed of Harry's model train is 0.7 m/s. The track is 35 m long.

How long does it take the train to complete a circuit of the track?

Time = $35 \div 0.7 = 50$ seconds.

C Rates

1. Rates, where quantities other than distance are covered in a certain time, are very similar to speed.
2. *Example*:

   In a factory, eggs can be packed at a rate of 1200 boxes per hour.

   How long does it take to pack 9000 boxes?

   In this example, the rate corresponds to speed and the number of boxes to distance.

   $$\text{time} = \frac{\text{distance}}{\text{speed}} = \frac{9000}{1200} = \frac{90}{12} = \frac{15}{2} = 7\frac{1}{2} \text{ hours.}$$

D Density

1. Density is another compound measure. When you say a given amount of one substance is 'heavier' than another, you are really talking about density.

>> **key fact** **Density** $= \dfrac{\textbf{mass}}{\textbf{volume}}$

Units of density include kilograms per cubic metre (kg/m^3) and grams per cubic centimetre (g/cm^3).

Density calculations are similar to speed calculations.
Density corresponds to speed, mass to distance and volume to time.

2. *Example*:

   200 cm^3 of potassium weighs 170 g.
   500 cm^3 of sodium weighs 480 g.

   Which element is denser?

   Potassium: density $= \frac{170}{200} = 0.85\ g/cm^3$.

   Sodium: density $= \frac{480}{500} = 0.96\ g/cm^3$.

   Sodium is denser than potassium.

>> practice questions

1 Calculate the missing value for each of these journeys.

	distance	time	average speed
(a)	30 km	2 hours	(km/h)
(b)	4 km	30 minutes	(km/h)
(c)	100 km	(hours)	40 km/h
(d)	(metres)	0.05 seconds	250 m/s
(e)	(km)	1 hour	8 km/second
(f)	400 m	1 minute	(km/h)

2 The density of platinum is 21.5 g/cm^3. That of gold is 19.3 g/cm^3.

Which is heavier, 50 cm^3 of platinum or 60 cm^3 of gold, and by how much?

3 In the UK, we consume about 80 million tonnes of oil per year. Estimate the rate of consumption in tonnes per second, correct to 2 sf.

Angle facts

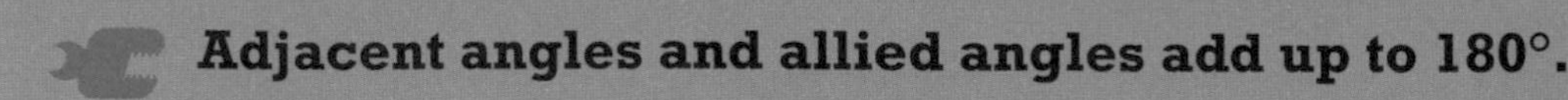

- **Adjacent angles and allied angles add up to 180°.**

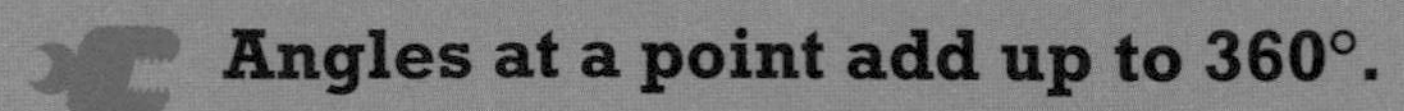

- **Angles at a point add up to 360°.**

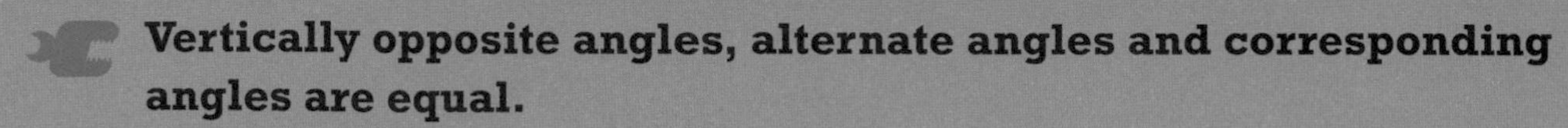

- **Vertically opposite angles, alternate angles and corresponding angles are equal.**

A Adjacent angles

1 **key fact** **Angles that are adjacent to each other on a straight line are supplementary – that is, they add up to 180°.**

In this diagram, $x + y = 180°$.

This is also true if there are three or more angles.

2 **key fact** **Angles at a point add up to 360°.**

In this diagram,
$p + q + r + s + t = 360°$.

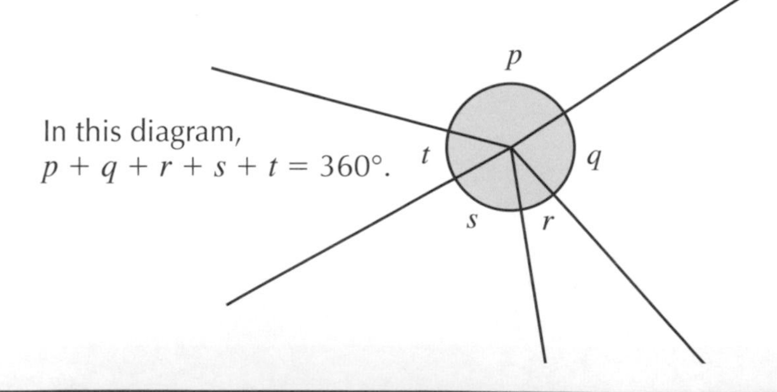

3 **key fact** **Vertically opposite angles are equal.**

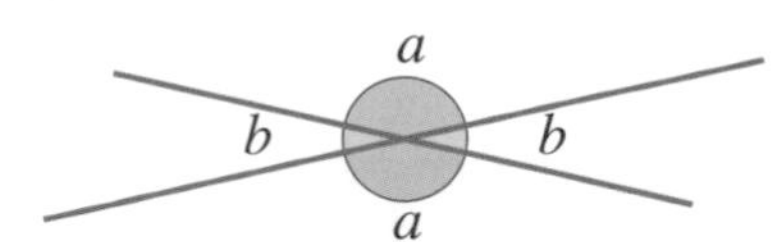

In this diagram, the two angles labelled a are equal and so are the two labelled b.

Vertically opposite angles are sometimes known as X-angles because of the shape of the diagram. Remember: 'vertically' means 'across a vertex', not 'upright'.

B Angles and parallel lines

1 Whenever sets of parallel lines are crossed by another straight line (called a **transversal** line), many angle relationships are formed.

2 **key fact** **Alternate angles are equal.**

You can remember this by thinking of the Z shape formed, as shown in the diagram. Alternate angles are sometimes known as Z-angles.

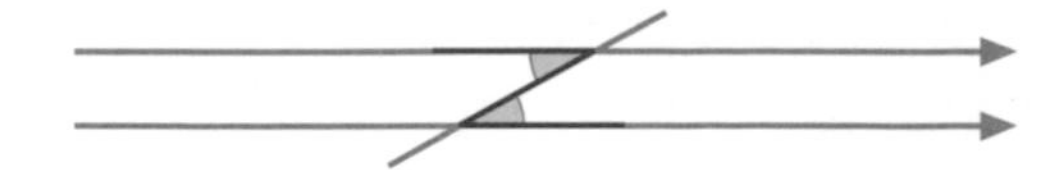

3 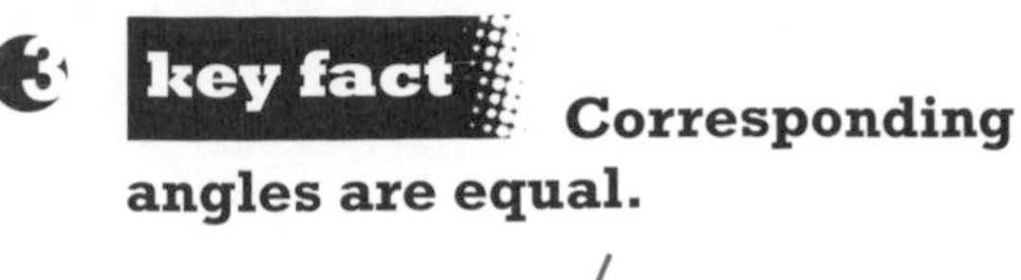

key fact **Corresponding angles are equal.**

You can remember this by thinking of the F shape formed, as shown in the diagram. Corresponding angles are sometimes known as F-angles.

4 

key fact **Allied angles are supplementary (add up to 180°).**

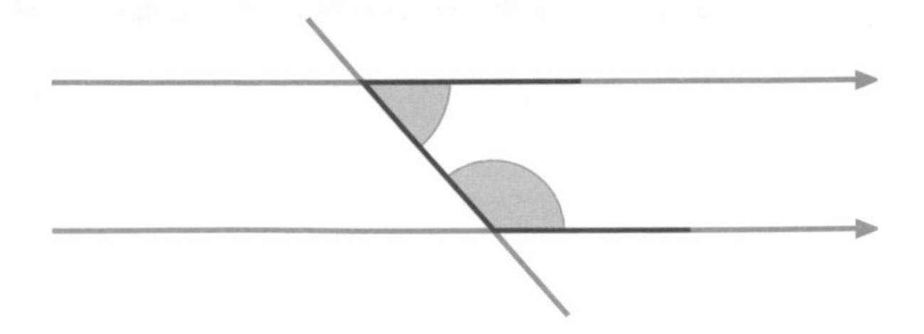

You can remember this by thinking of the C shape formed, as shown in the diagram. Allied angles are sometimes known as C-angles.

5 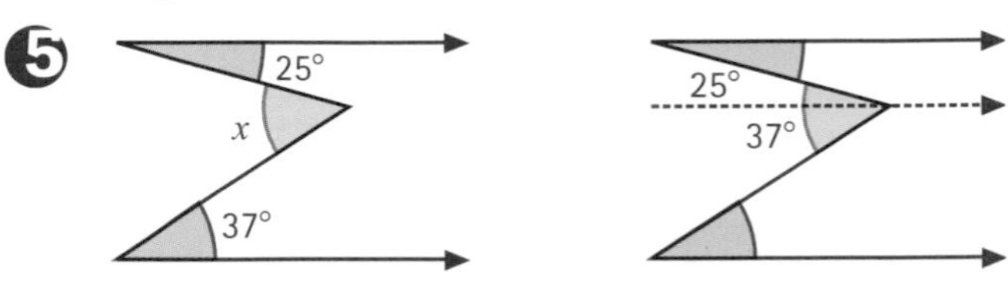

In a situation like this, you need to add an extra parallel line to the diagram. The angle x that you are trying to find is just made up of $25° + 37° = 62°$, because of the alternate angles created by the new parallel line.

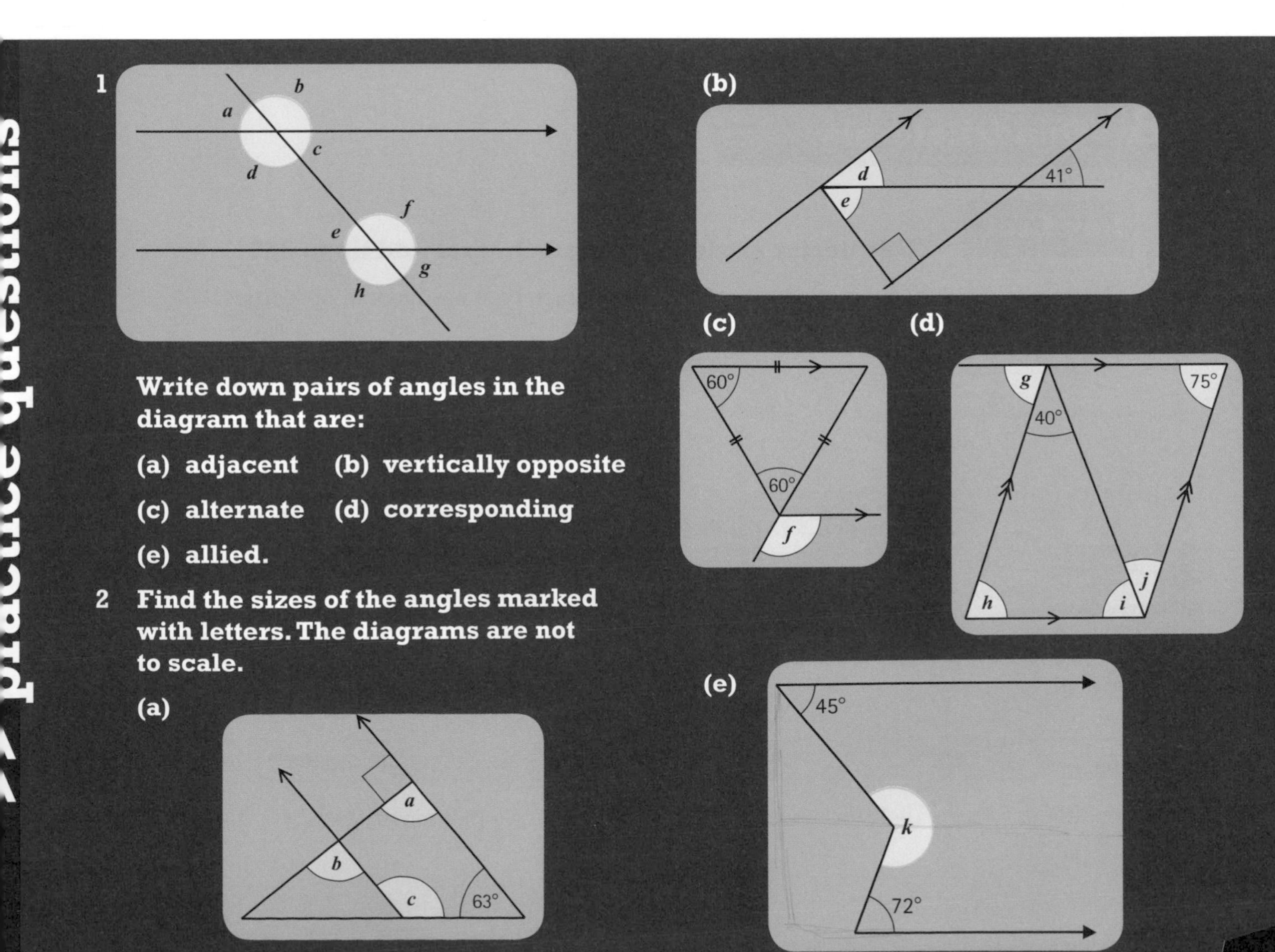

1 **Write down pairs of angles in the diagram that are:**

(a) adjacent (b) vertically opposite

(c) alternate (d) corresponding

(e) allied.

2 **Find the sizes of the angles marked with letters. The diagrams are not to scale.**

(a)

(b)

(c)

(d)

(e)

Properties of shapes

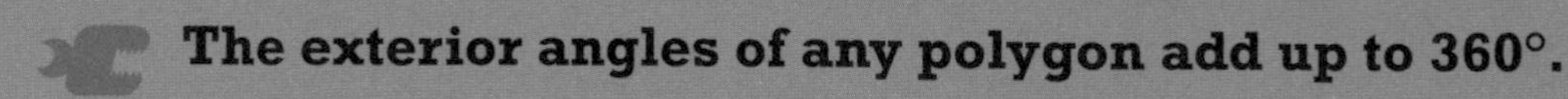

- **The exterior angles of any polygon add up to 360°.**
- **The angle sum of any polygon is a multiple of 180°.**
- **Regular polygons have all sides and angles equal.**

A Triangles

1 The angles inside a triangle at the vertices (corners) are called its **interior** angles.

>> **key fact** **The interior angles of a triangle add up to 180°.**

You make an **exterior** angle by extending one of the sides. In this situation, $x + y = z$.

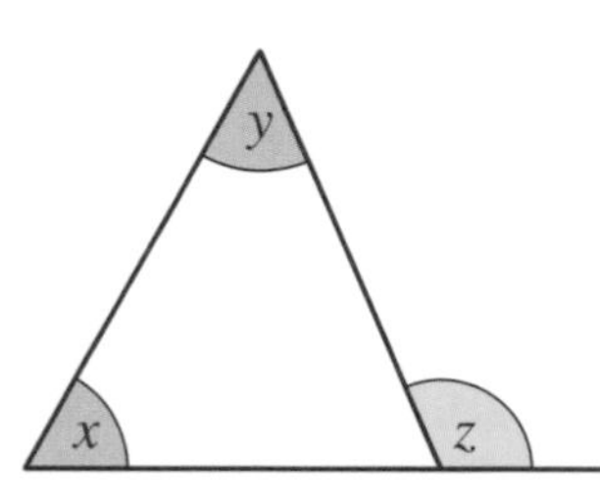

2 **Isosceles** triangles have two sides of equal length and two equal angles. You can use this fact when finding unknown angles.

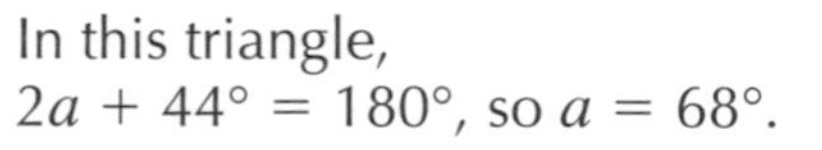

In this triangle, $2a + 44° = 180°$, so $a = 68°$.

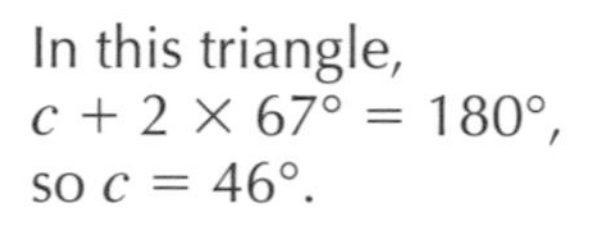

In this triangle, $c + 2 \times 67° = 180°$, so $c = 46°$.

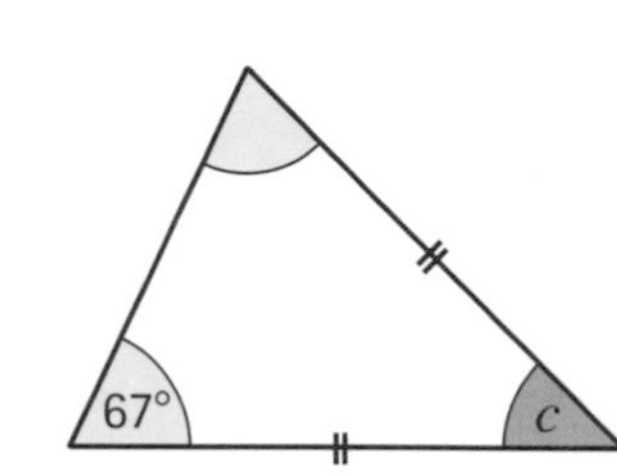

3 All three sides of an **equilateral** triangle are equal and all its angles are 60°.

B Quadrilaterals

1 **key fact** **The interior angles of a quadrilateral add up to 360°.**

There are many different types of quadrilateral. Each type has its own properties.

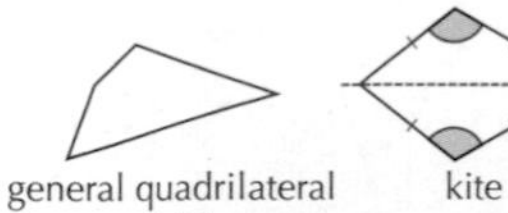
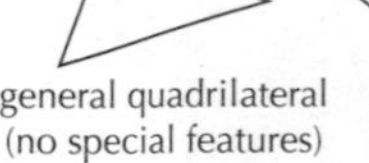
general quadrilateral (no special features)

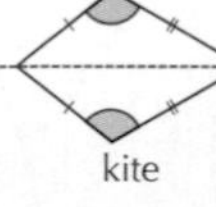
kite

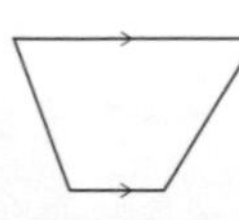
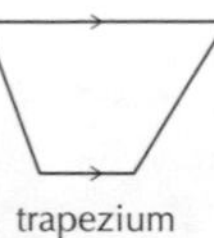
trapezium

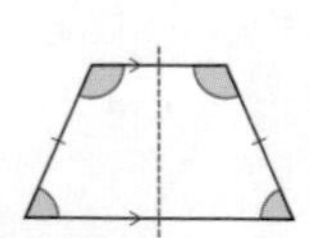
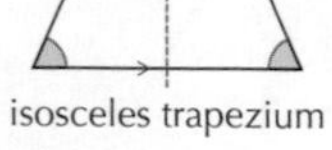
isosceles trapezium

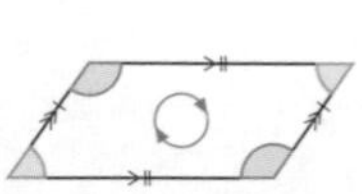
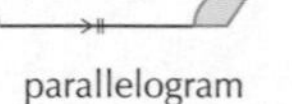
parallelogram

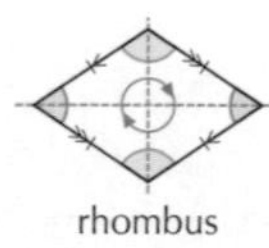
rhombus

rectangle

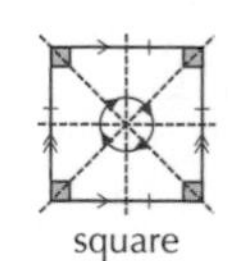
square

	kite	trapezium	parallelogram	rhombus	rectangle	square
opposite sides	–	1 ‖ pr	2 ‖ pr, 2 = pr	2 ‖ pr, all =	2 ‖ pr, 2 = pr	2 ‖ pr, all =
adjacent sides	2 = pr	–	–	all =	⊥	= and ⊥
opposite angles	1 = pr	–	2 = pr	2 = pr	all = (90°)	all = (90°)
adjacent angles	–	2 supp pr	2 supp pr	2 supp pr	all = (90°)	all = (90°)
diagonals	⊥	–	=	= and ⊥	=	= and ⊥
symmetry	1 line	–	order 2	2 lines, order 2	2 lines, order 2	4 lines, order 4

Key: equal (=), parallel (‖), perpendicular (⊥), pair (pr), supplementary (supp).

C Polygons

1 The exterior angles of a polygon always add up to 360°.
One interior angle and its exterior angle always add up to 180°.

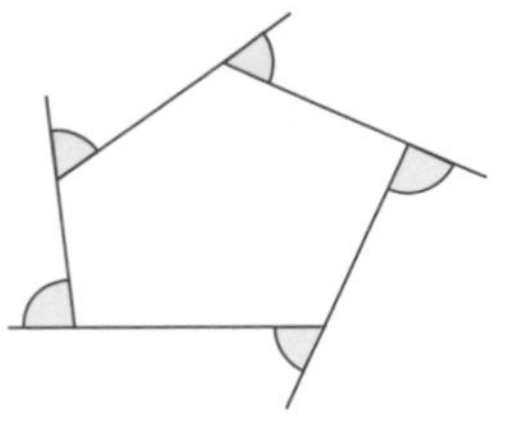

This table gives the names of different types of polygon.

Number of sides	5	6	7	8	9	10	12
Name	pentagon	hexagon	heptagon	octagon	nonagon	decagon	dodecagon

2 The angle sum of a polygon is the total of all its **interior** angles. A polygon with n sides has an angle sum of $180(n - 2)°$, or $180n - 360°$. If you split the polygon up into triangles, you can see why.

3 This table gives angle sums for some of the polygons.

Number of sides	3	4	5	6	7	8	10
Angle sum	180°	360°	540°	720°	900°	1080°	1440°

D Regular polygons

1 **key fact** **In a regular polygon, all the sides are the same length, all the interior angles are equal and all the exterior angles are equal.**

In this section, n stands for the number of sides.

2 As all the exterior angles add up to 360°, one exterior angle = $\frac{360°}{n}$, and so one interior angle is $180° - \frac{360°}{n}$.

3 You can use this to find out how many sides a regular polygon has, if you know one of its interior angles. Suppose one interior angle is 150°. Then the exterior angle = 180° − 150° = 30°.

So $\frac{360°}{n} = 30°$ and $n = 12$.

The shape is a dodecagon.

This table gives results for some of the polygons.

Number of sides	3	4	5	6	8	10	12
Exterior angle	120°	90°	72°	60°	45°	36°	30°
Interior angle	60°	90°	108°	120°	135°	144°	150°

>> practice questions

1 Find the angle marked x in each diagram.

(a)

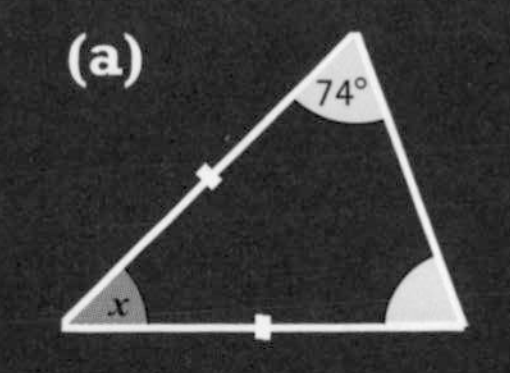

(b)

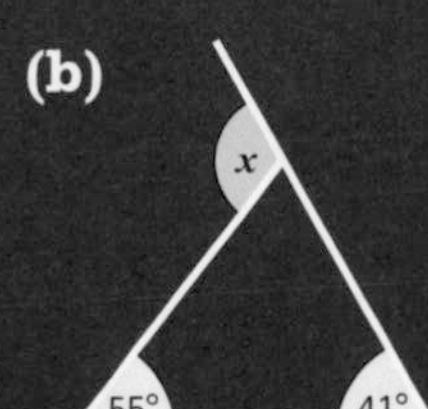

(c)

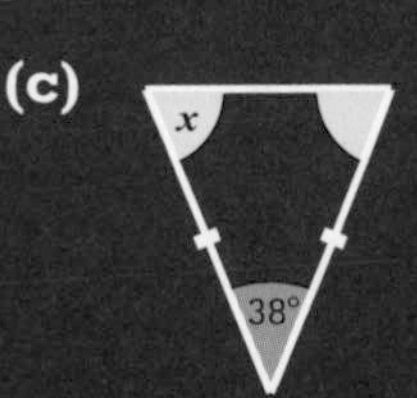

(d)

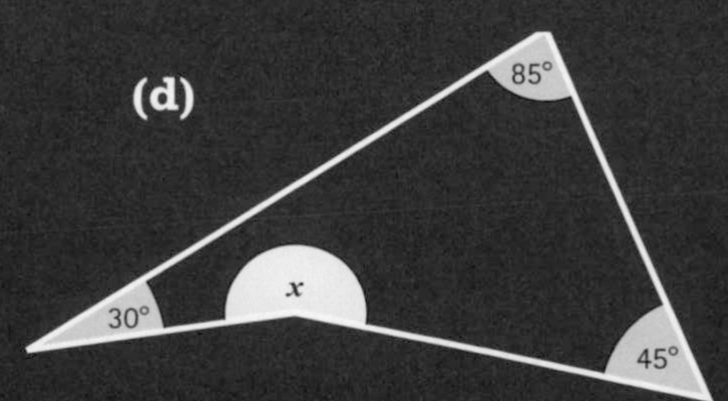

2 Calculate the angle sum for: (a) a dodecagon (12 sides) (b) an icosagon (20 sides).

3 The angle sum of a regular polygon is 1260°. Calculate:

(a) the size of one interior angle (b) the size of one exterior angle
(c) the number of sides.

Pythagoras' Rule

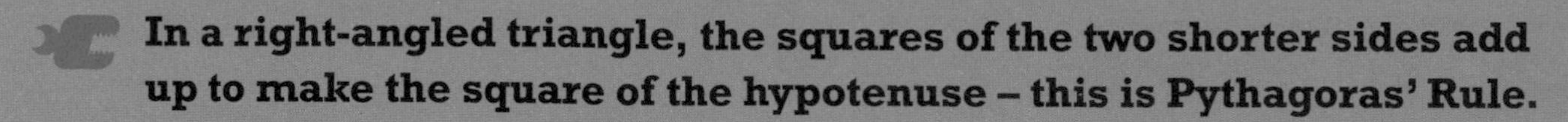

- **In a right-angled triangle, the squares of the two shorter sides add up to make the square of the hypotenuse – this is Pythagoras' Rule.**
- **To find one of the shorter sides, square and subtract.**

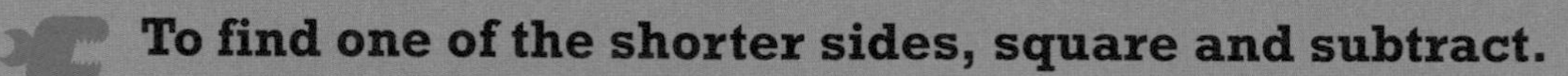

A Pythagoras' Rule

1 **key fact** **The hypotenuse in a right-angled triangle is its longest side. It is always opposite the right angle.**

In this diagram, the hypotenuse is labelled h and the two shorter sides that form the right angle, a and b. (It doesn't matter which way round you label a and b.)

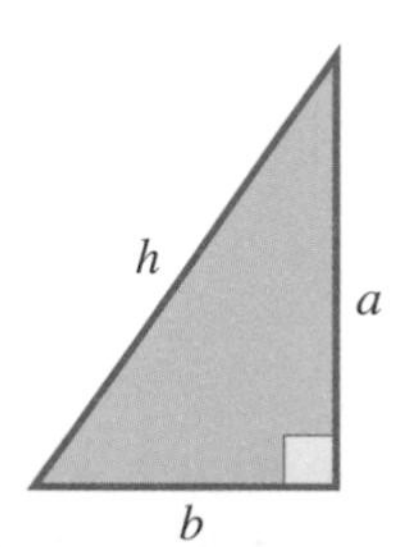

>> **key fact** **Pythagoras' Rule states that $h^2 = a^2 + b^2$.**

In words, 'the squares of the two shorter sides add up to make the square of the hypotenuse'.

B Finding the hypotenuse

1 You can use this to find the hypotenuse of any right-angled triangle.

In this triangle, a = 4 cm and b = 3 cm.

h
4 cm
3 cm

By Pythagoras' Rule,

$h2 = a^2 + b^2$
$= 4^2 + 3^2$ *substitute given values*
$= 16 + 9$ *calculate the squares*
$= 25$ *simplify*

So $h = \sqrt{25}$ = **5 cm**.

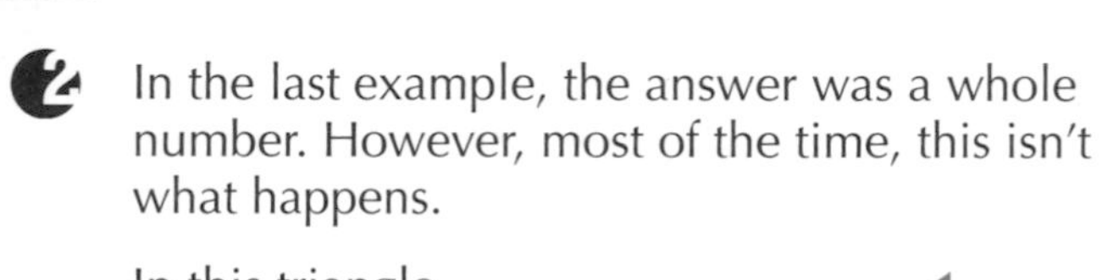

2 In the last example, the answer was a whole number. However, most of the time, this isn't what happens.

In this triangle, a = 5 cm and b = 7 cm.

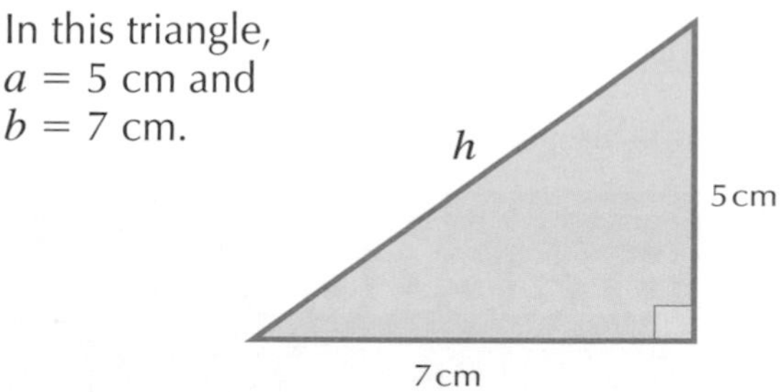

By Pythagoras' Rule,

$h^2 = a^2 + b^2$
$= 5^2 + 7^2$ *substitute given values*
$= 25 + 49$ *calculate the squares*
$= 74$ *simplify*

So $h = \sqrt{74}$ = **8.60 cm** (2 dp).

This answer has to be rounded to a suitable degree of accuracy.

C Finding the shorter side

1. Just apply the rule as before – the only difference is that you will need to subtract squares.

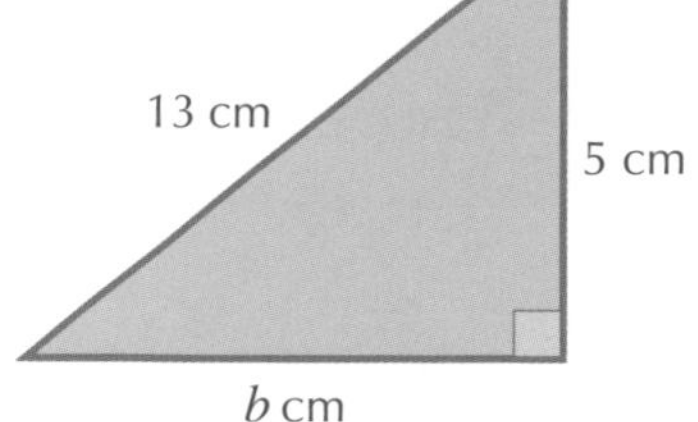

By Pythagoras' Rule, $h^2 = a^2 + b^2$

So $13^2 = 5^2 + b^2$ *substitute given values*

So $169 = 25 + b^2$ *calculate the squares*

So $169 - 25 = b^2$ *rearrange the equation*

So $144 = b^2$ simplify

So $b = \sqrt{144} =$ **12 cm**.

D Applying the Rule

1. You can find the length of the diagonal of a rectangle. The diagonal splits the rectangle into two congruent right-angled triangles.

Example:

Find the diagonal (d) of a rectangle that is 8 cm by 5 cm.

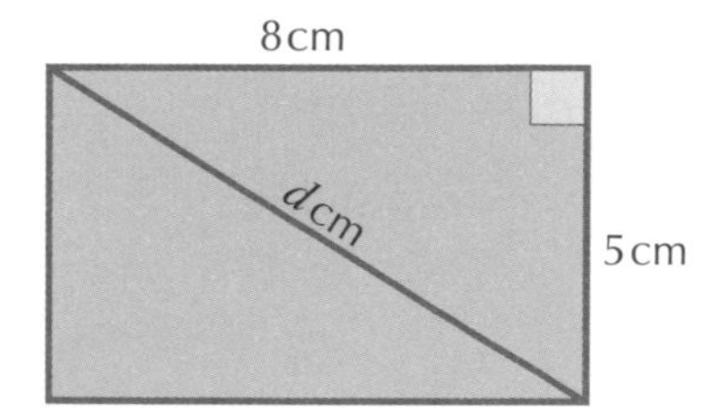

By Pythagoras' Rule,

$$d^2 = 5^2 + 8^2$$
$$= 25 + 64 = 89$$

So $d = \sqrt{89} =$ **9.43 cm** (2 dp).

2. You can find the distance between two points on a co-ordinate grid using Pythagoras' Rule.

Example:

Find the distance between the points $P(0, -5)$ and $Q(6, 3)$.

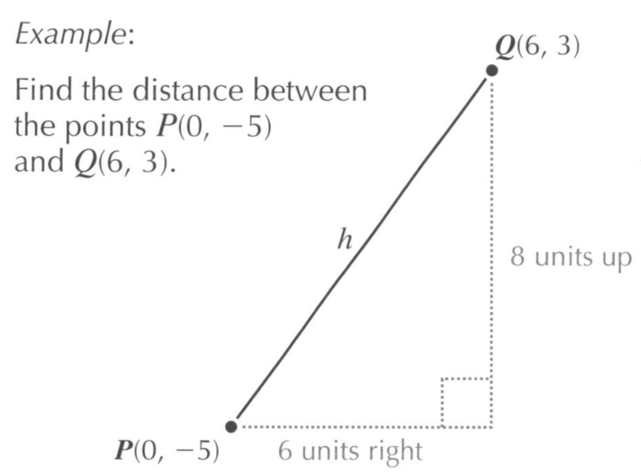

By Pythagoras' Rule,

$$h^2 = 8^2 + 6^2$$
$$= 64 + 36 = 100$$

So $h = \sqrt{100} =$ **10 units**.

>> practice questions

1 Find x in the following triangles.

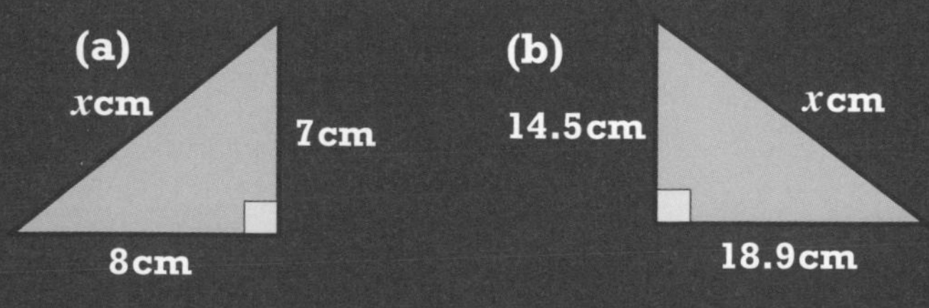

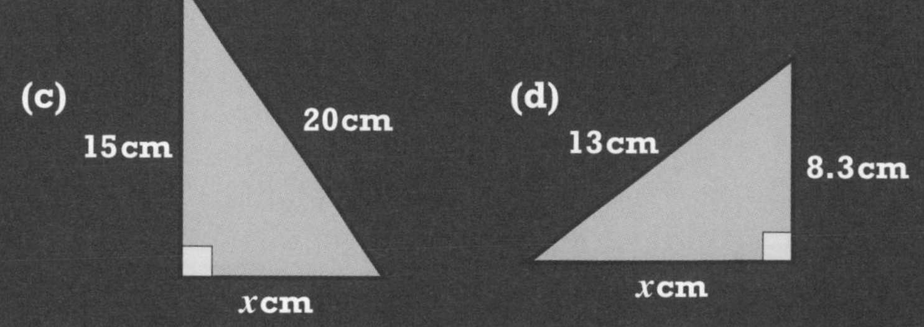

2 A 5 m ladder rests against a wall with its foot 2 m from the wall. How far up the wall does the ladder reach?

3 A piece of curtain material is a rectangle of length 2.5 m and width 1.6 m. How long is the diagonal?

4 Find the distance between each pair of co-ordinate points.

(a) (0, 7) and (7, 0) (b) (1, 4) and (8, 8)

5 Martha and Prakash were arguing about a triangle that had sides of 9 cm, 12 cm and 15 cm. Prakash said the triangle had a right angle. Martha said it did not. Who was correct?

...culating areas

- Area is the amount of surface covered by a 2-dimensional shape.
- Most simple shapes have formulae that can be used to calculate their areas.
- More complicated (compound) shapes can be built from simple ones.

A Rectangles and squares

>> **key fact** **The area of a rectangle = length × width (lw).**
The area of a square = length × length (l^2).

1. In this rectangle $l = 100$ and $w = 25$.

100 m

25 m

$A = lw$

$A = 25 \times 100$

$A = 2500\,\text{m}^2$

2. Compound shapes can be made by **putting rectangles together**, or **removing one rectangle from another**.

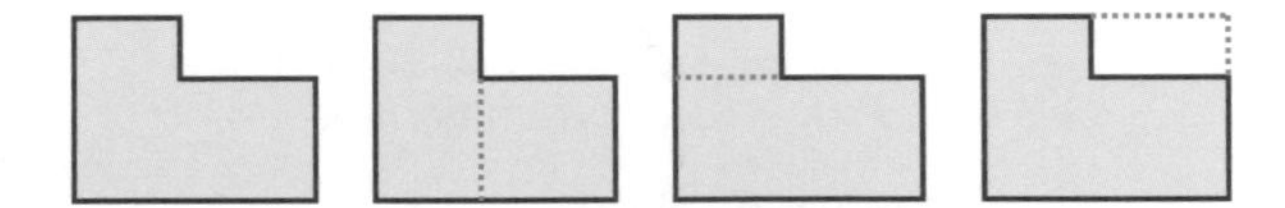

B Triangles

1. **key fact** **The area of a triangle = $\frac{1}{2}$ bh.**

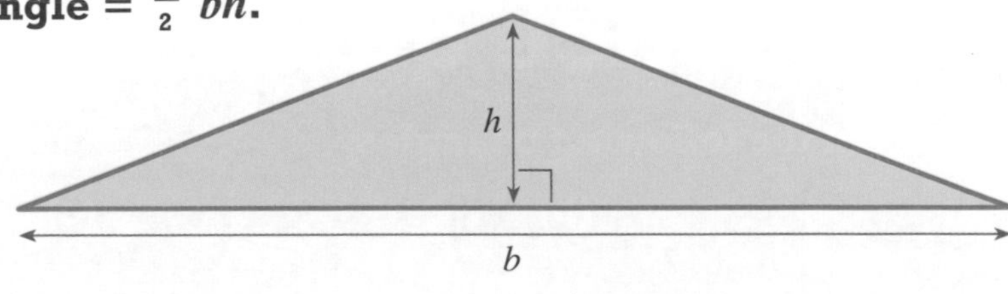

b is the base length and h is the perpendicular height.

The **perpendicular height** is the line that forms a **right angle** with the base and passes through the **third vertex** of the triangle.

2. Find the area of a triangle of base length 5 cm and height 12 cm.

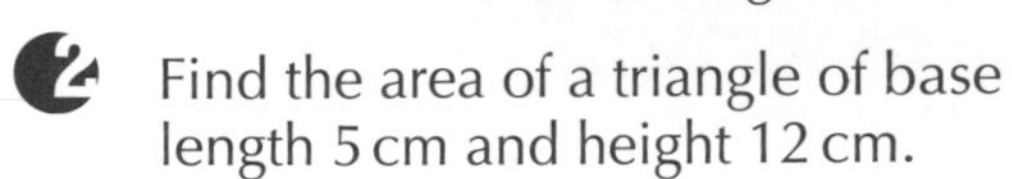

$A = \frac{1}{2} \times 5 \times 12$

$= 30\,\text{cm}^2$

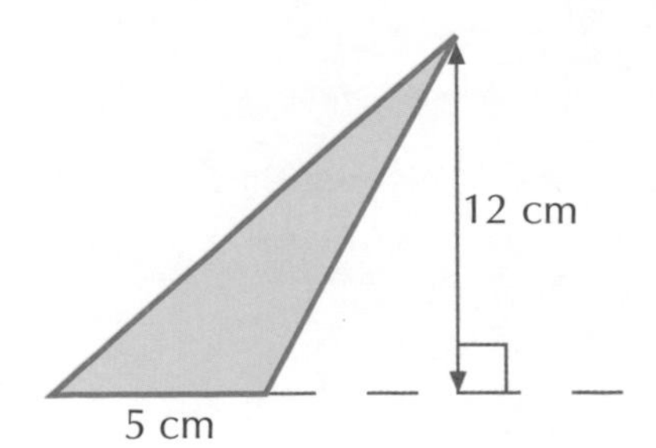

C Parallelograms and trapezia

1 Parallelograms and trapezia both use the perpendicular height in their area formulae:

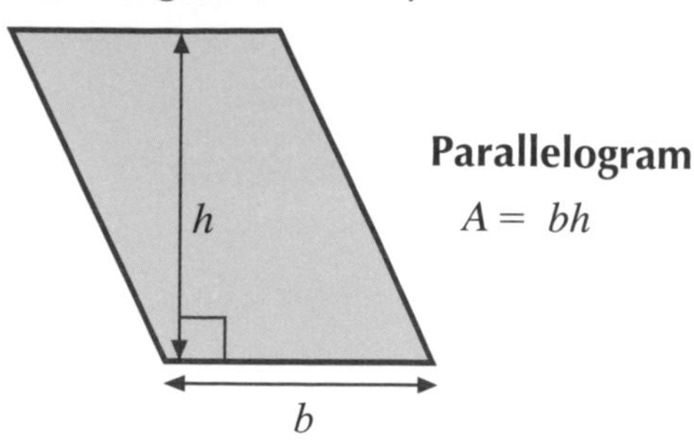

Parallelogram

$A = bh$

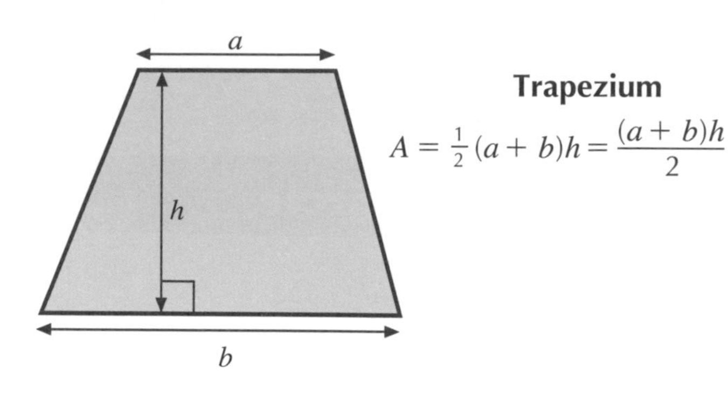

Trapezium

$A = \frac{1}{2}(a + b)h = \frac{(a + b)h}{2}$

2 Calculate the area of this trapezium:

The perpendicular height isn't given, so you will have to calculate it using Pythagoras' Rule and the triangle on the right.

$$h^2 = 10^2 - 6^2$$
$$= 100 - 36 = 64$$
So $h = 8\,\text{cm}$

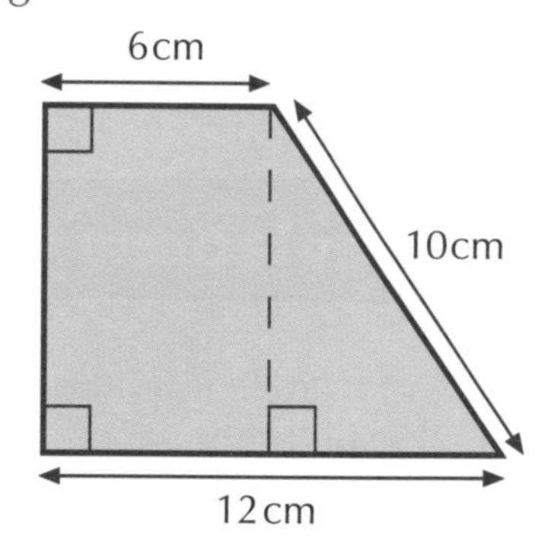

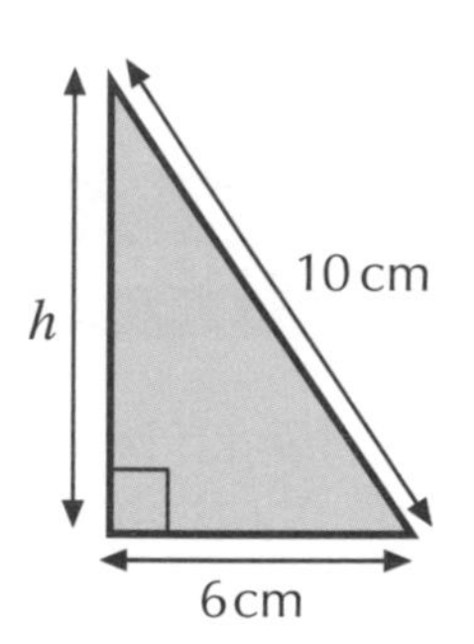

Using the trapezium formula with $a = 6\,\text{cm}$, $b = 12\,\text{cm}$, and $h = 8\,\text{cm}$ gives:

$A = \frac{1}{2}(a + b)h = 72\,\text{cm}^2$.

>> practice questions

1 **Find the area of this rectangle.**

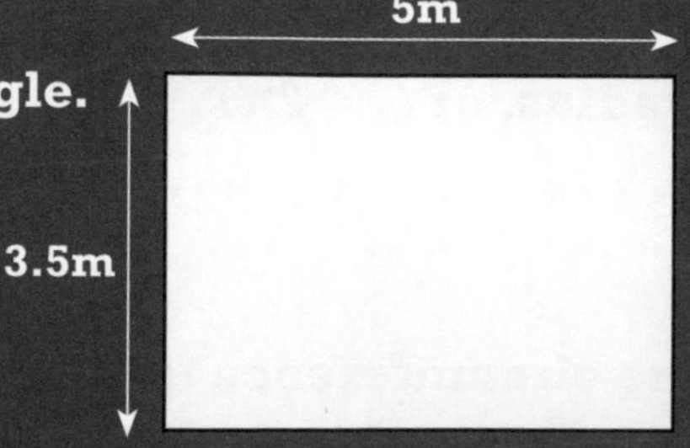

2 **Find the area of a carpet needed for this office layout.**

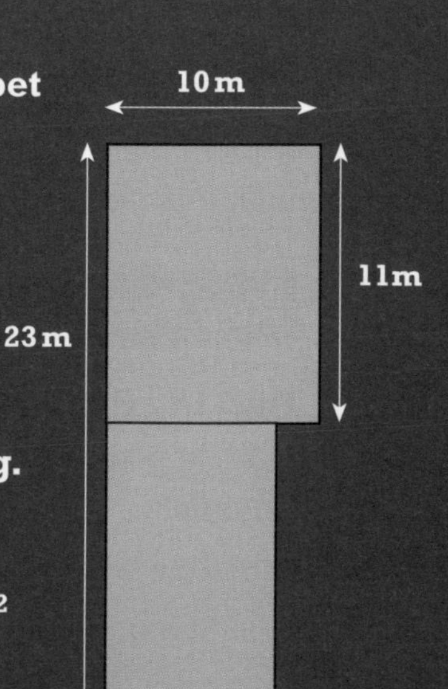

3 **Catherine has to paper a bedroom wall that is 3 m high and 10 m long. The wallpaper she wants comes in rolls 2 m wide and 5 m long. How many rolls of paper will she need?**

4 **Martin wants to paint a fence. The fence is 14 m long and 3 m high. He has to paint both sides of the fence. One tin of paint covers 12 m^2 of fence. How many tins will he need?**

5 **Find the area of a triangle that is 6.5 cm in width and 7 cm in height.**

6 **A trapezium has parallel sides of length 5 cm and 11 cm. Its area is 80 cm^2. What is its perpendicular height?**

Circle calculations

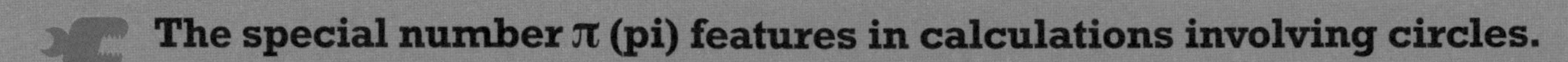
The special number π (pi) features in calculations involving circles.

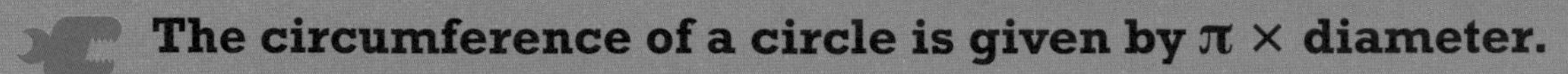
The circumference of a circle is given by π × diameter.

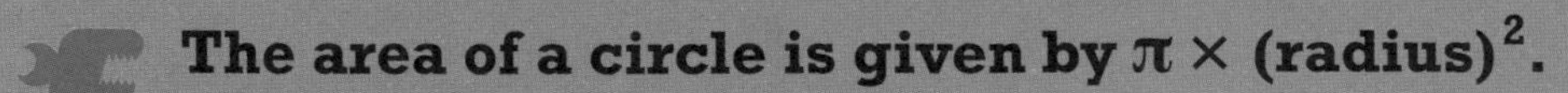
The area of a circle is given by π × (radius)².

A Parts of a circle

1. The diagrams give the names of parts of a circle.

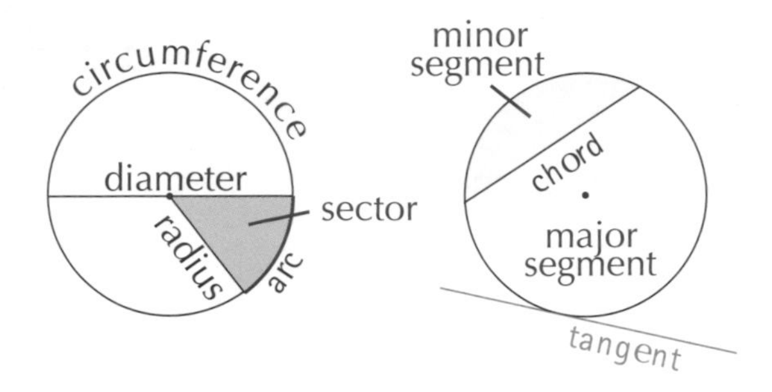

B Circumference calculations

1. The length of the circumference of a circle is a little more than 3 times its diameter. The actual number is about 3.14 and is called **pi** (pronounced 'pie'). This is a letter from the Greek alphabet and is written π (a bit like a T, but with two vertical lines). You should have a π key on your calculator. If not, use an approximate decimal value such as 3.14 or 3.142.

2. **key fact** **Circumference = π × diameter, or $C = \pi d$.**

   So if a circle is 5 cm across, its circumference is $C = \pi d = \pi \times 5 = 15.71$ cm (2 dp).

3. If you are given the radius, you can double it to find the diameter and then use $C = \pi d$. Alternatively, you can use the other circumference formula:

>> **key fact** **Circumference = 2 × π × radius, or $C = 2\pi r$.**

   So if a circle's radius is 10 m, its circumference is $C = 2\pi r = 2 \times \pi \times 10 = 62.83$ m (2 dp).

4. **key fact** **To work backwards from the circumference to find the diameter, use $d = \frac{C}{\pi}$.**

   Example:

   A trundle wheel's circumference is exactly 1 m. What is its diameter?

   $d = \frac{1}{\pi} = 0.318$ m $= 31.8$ cm (3 sf).

C Area calculations

1 The number π also appears when you calculate the area of a circle.

>> **key fact** **Area of a circle = π × the square of the radius, or $A = \pi r^2$.**

2 The circle from section B2 with diameter 5 cm has radius 2.5 cm.

So its area is
$A = \pi r^2 = \pi \times 2.5^2 = 19.63\text{ cm}^2$ (2 dp).

3 Working backwards from the area to the radius is a little harder: $r = \sqrt{\frac{A}{\pi}}$.

So a circular pond covering an area of 4 m^2 would have radius $r = \sqrt{\frac{4}{\pi}} = 1.13\text{ m}$.

D Leaving π in the answer

1 In some exam questions, you may be asked to write an answer 'in terms of π'. This means that, instead of calculating a decimal answer using the π button on your calculator, you simply write down what multiple of π is involved. You can then easily compare two answers.

2 *Example*:

Circle A has diameter 20 cm. Circle B has radius 20 cm. Write the areas of the two circles in terms of π, and find how many times larger the area of B is than that of A.

Radius of $A = 20 \div 2 = 10$ cm.

Area of $A = \pi \times 10^2 = 100\pi$.
Area of $B = \pi \times 20^2 = 400\pi$.

The area of A is 4 times the area of B.

>> practice questions

1 Find the circumference of each of the following circles, using the value for π on your calculator or 3.14:

(a) radius 7 cm **(b) diameter 4 cm**
(c) diameter 25 cm **(d) radius 10.9 cm**

2 Find the areas of the following circles:

(a) **(b)**

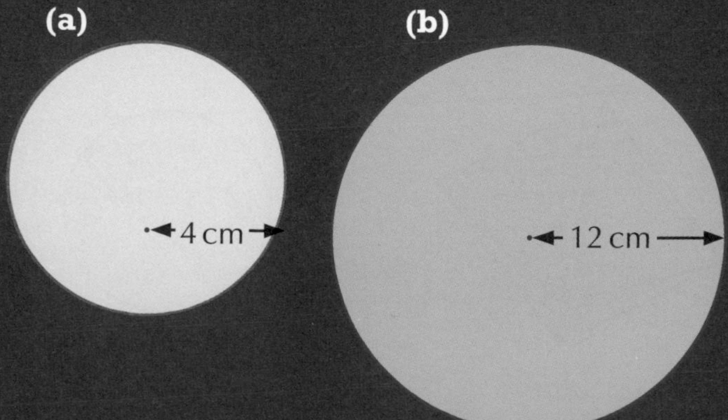

3 A circular pond has a diameter of 4.5 m.

It is surrounded by a path that is 1.2 m wide.

Find: (a) the area of the pond

(b) the area of the path.

4 A circle has an area of 100 cm^2; find the radius.

5 Circle $C1$ has diameter 4 cm. Circle $C2$ has diameter 12 cm.

(a) Express the perimeter and area of each circle in terms of π.

(b) How many times longer is $C2$'s circumference than $C1$'s?

(c) How many times more area does $C2$ cover than $C1$?

Solid shapes

Solid (3D) shapes have faces, vertices and edges.

3D shapes can be represented on paper using isometric drawing, a net, or a plan and elevations.

A Shapes in 3D

Three-dimensional (3D) shapes have faces, edges and vertices.

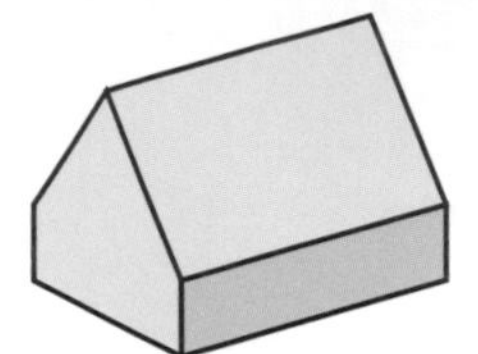
This prism has ...

... 7 faces,

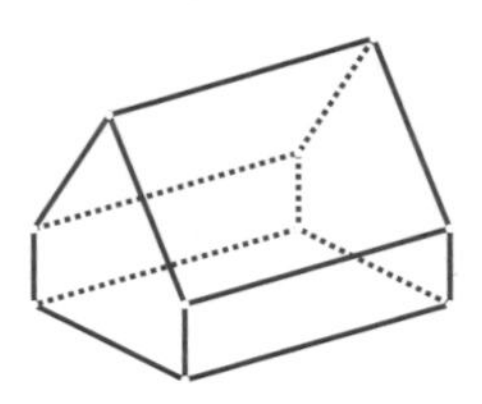
...15 edges,

... 10 vertices

Shapes with flat faces and straight edges are called polyhedra.

Prisms have the same cross-section all the way through. This is sometimes called having a **uniform** cross-section. These are all prisms.

hexagonal prism

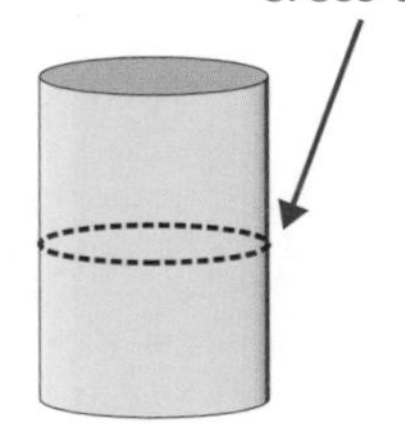

cylinder

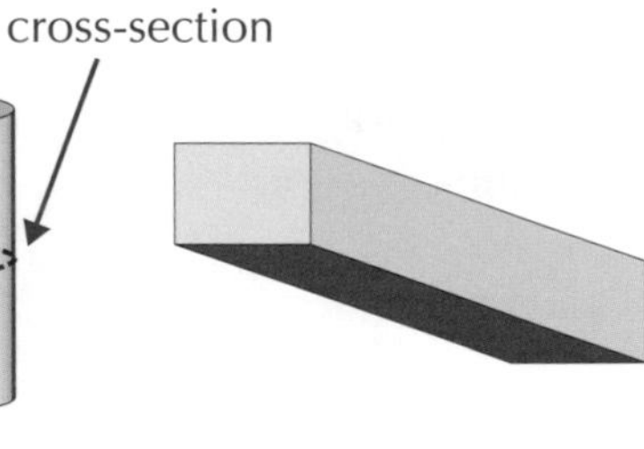
cuboid

A **pyramid** has a flat base joined to an apex. These are all pyramids.

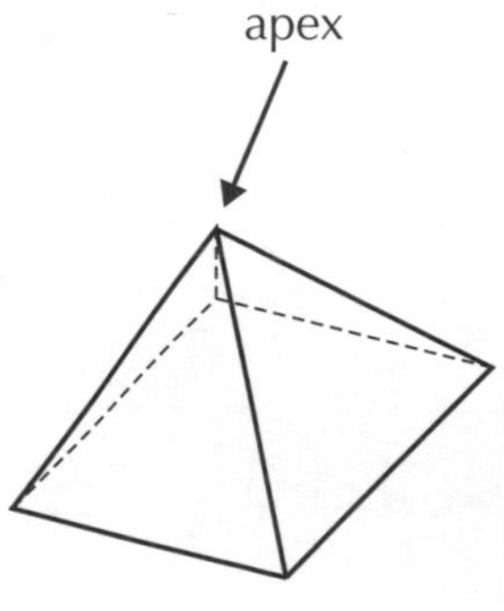

square-based pyramid

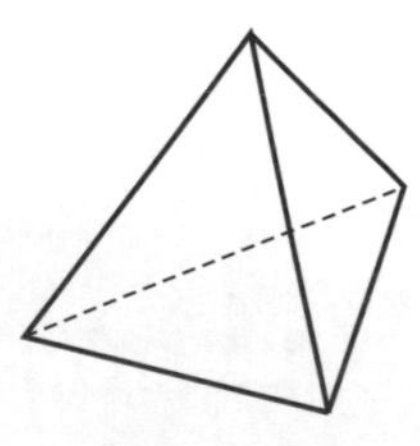
tetrahedron

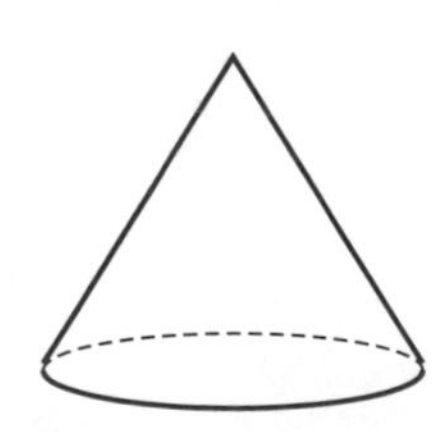
cone

These are examples of other types of solid shape.

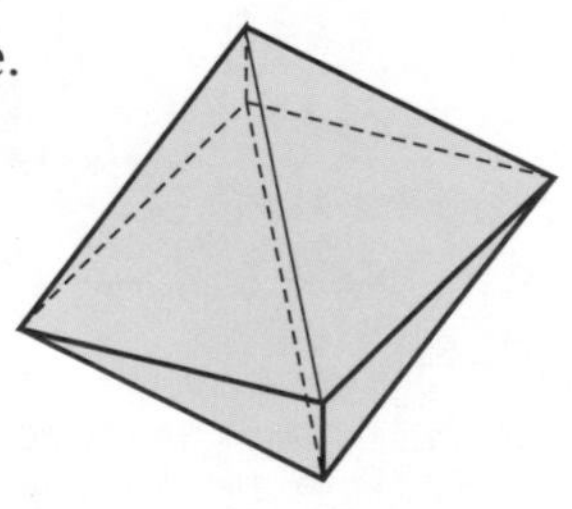
octahedron

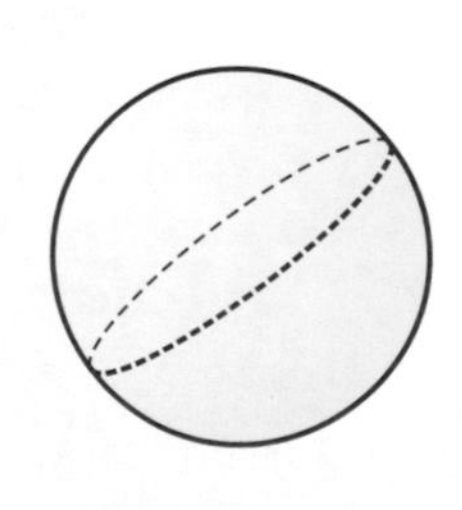
sphere

B Isometric drawings

1. One way of making a realistic drawing of a solid shape is to use **isometric** paper.

   Mark the measurements of the shape on your drawing if possible.

2. You might be given a description of the shape to draw, or a set of plans and elevations (see section D).

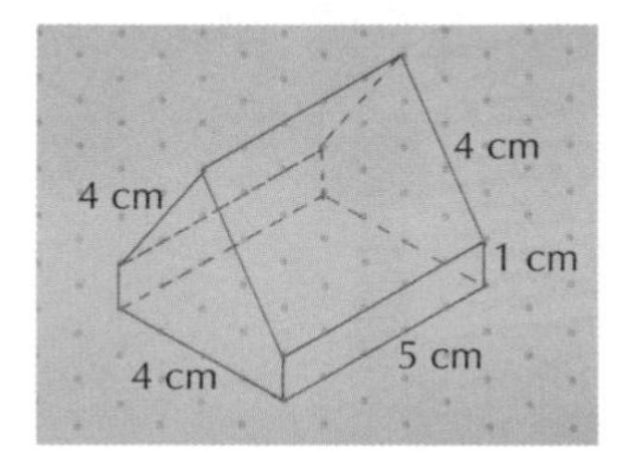

C Nets

1.  **key fact** **The net of a shape is a 'map' of all the faces laid out flat.**

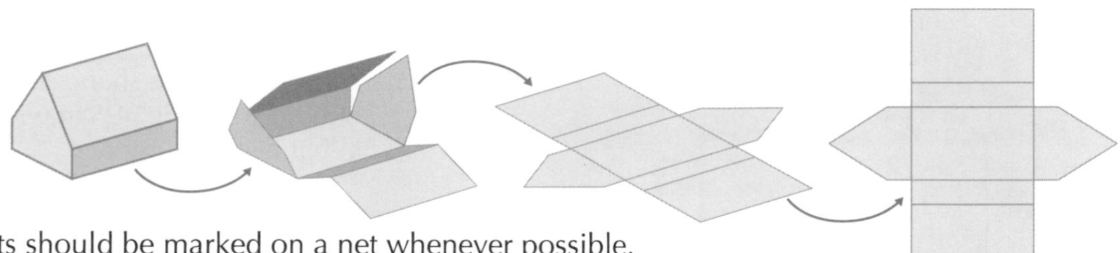

Measurements should be marked on a net whenever possible.

The area of all the faces added together is equal to the **surface area** of the shape.

D Plans and elevations

1. Another way of representing a 3D shape on paper is with a **plan and elevations**.

2. **key fact** **Plans and elevations show no perspective.**

   They are like 'maps' – or photos taken from a great distance and magnified. You should always mark measurements on plans and elevations.

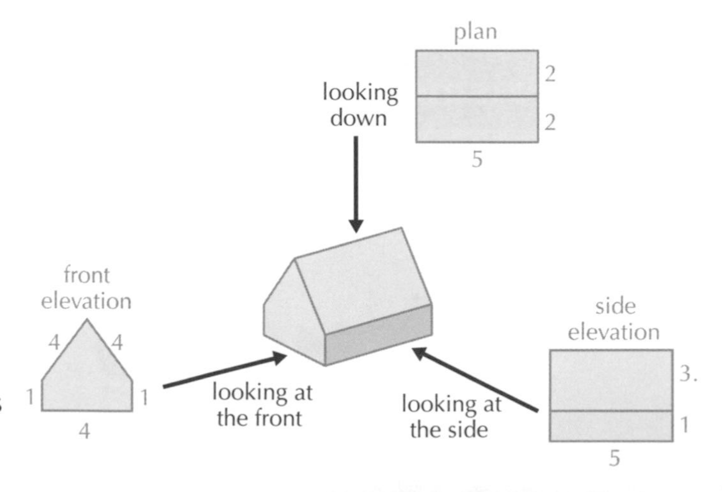

>> practice questions

1 **For each of the following shapes, (i) write down the number of faces, edges and vertices it has; (ii) draw its net; (iii) draw a plan and elevations.**

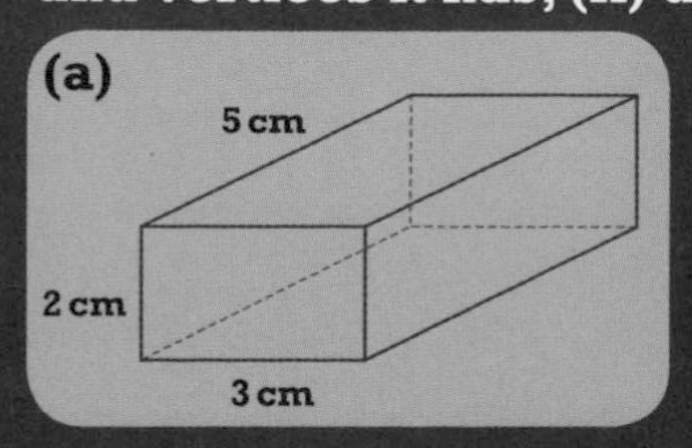

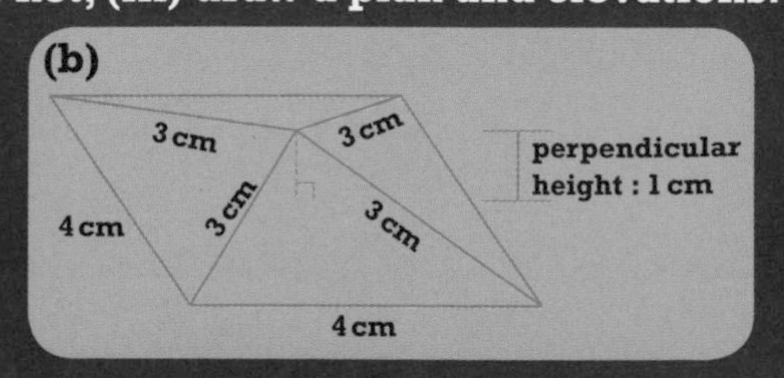

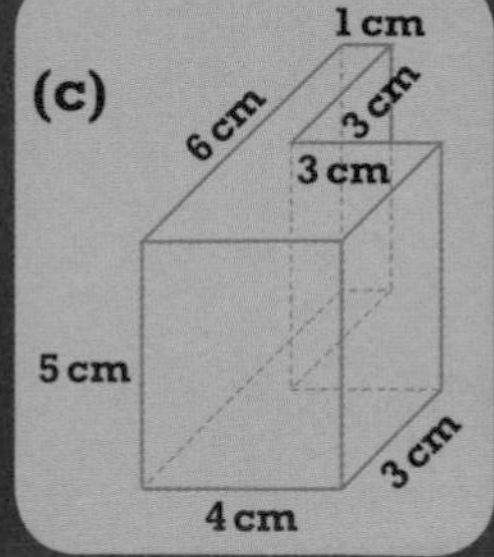

Volume calculations

- **A cuboid is a prism with a rectangular cross-section, so its volume = length × width × height.**
- **The volume of any prism is equal to the area of its cross-section multiplied by its length (or height).**
- **A cylinder is a circular prism.**

A Cuboids

1. **Volume** is the amount of three-dimensional space taken up by a solid shape.

2. 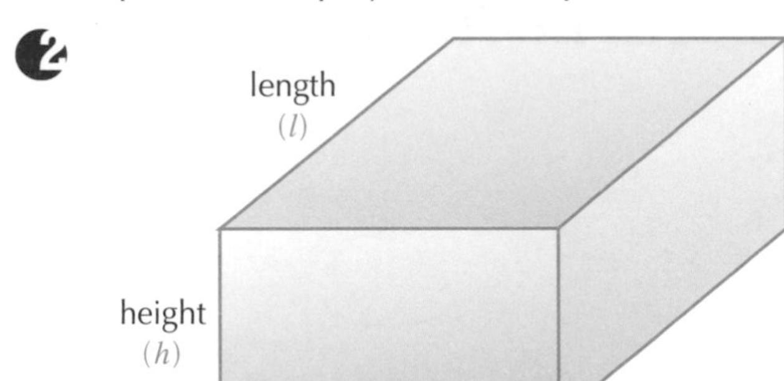

>> key fact **Volume of a cuboid = length × width × height; $V = lwh$.**

Example:

A room is in the shape of a cuboid 8 m long, 4.5 m wide and 3 m high. What is its volume?

$V = lwh = 8 \times 4.5 \times 3 = 108\,\text{m}^3$.

Remember: volume is measured in cubic units.

B Prisms

1. 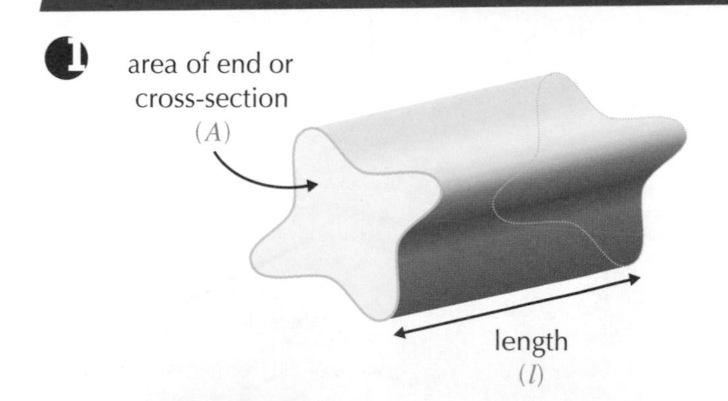

>> key fact **Volume of a prism = Area of cross-section × length; $V = Al$.**

Note that if a prism is 'standing on end', the length measurement may be thought of as a height, and the formula would be $V = Ah$.

2. *Example*:

Find the volume of this triangular prism.

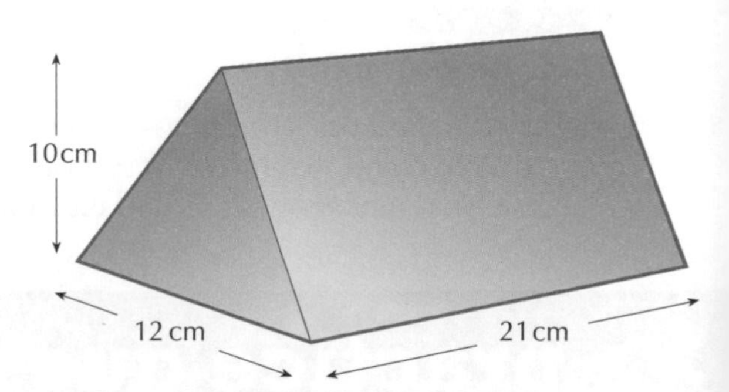

Area of cross section

$A = \frac{1}{2}bh = \frac{1}{2} \times 12 \times 10 = 60\,\text{cm}^2$.

Volume $V = Al = 60 \times 21 = 1260\,\text{cm}^3$.

C Cylinders

1 **key fact** **Cylinders are prisms with circular cross-sections. The volume formula is $V = Al = \pi r^2 l$.**

2 *Example*:

Find the capacity of this cylindrical can.

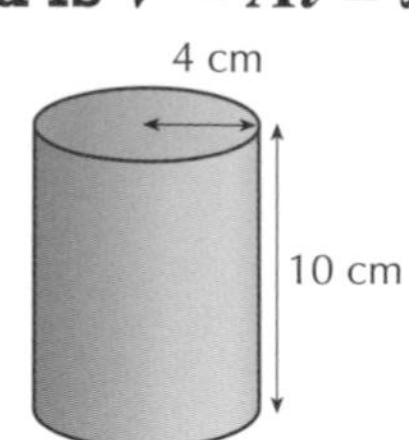

Area of cross-section
$A = \pi r^2 = \pi \times 4^2 = 16\pi$.

Volume of cylinder $V = Ah = 16\pi \times 10$
$= 503\text{ cm}^3$ (to nearest cm^3).

As $1\text{ cm}^3 = 1\text{ ml}$, the can will hold 503 ml or 0.503 l (to 3 sf).

Note that, if required, the answer could have been left in terms of π: 160π ml.

D 'Reverse' volume problems

1 In some kinds of problem, you may be given the volume of a shape and asked to find an unknown measurement.

2 *Example*: An oil tank is in the shape of a cuboid with a base 1.6 m by 0.8 m. 1000 litres of oil is to be poured into the tank. How deep will the oil be, to the nearest centimetre?

The oil will form a cuboid shape when poured in – the height is unknown.
The volume of the oil is 1000 litres = 1 m^3.

$V = lwh$
So $1 = 1.6 \times 0.8 \times h = 1.28 \times h$
So $h = 1 \div 1.28 = 0.78125\text{ m} = 78.125\text{ cm} = 78\text{ cm}$ (nearest cm).

3 *Example*: A cylindrical piece of wire is made from 10 cm^3 of copper. The wire is 50 m long. What is its diameter in millimetres, correct to 2 sf?

Work in cm, then convert to mm at the end: the wire is 50 × 100 = 5000 cm long.

$V = Al$
So $10 = A \times 5000$
So $A = 10 \div 5000 = 0.002\text{ cm}^2$
$A = \pi r^2$
So $0.002 = \pi r^2$
So $r^2 = 0.002 \div \pi$ and $r = \sqrt{0.002 \div \pi} = 0.02523\ldots\text{ cm}$

So diameter $= 2 \times 0.02523\ldots\text{ cm} = 0.05046\ldots\text{ cm} = 0.5046\ldots\text{ mm} = \mathbf{0.50\text{ mm}}$ (2 sf).

>> practice questions

1. **Find the volume of a cuboid of dimensions 2.5 m, 4 m and 6 m.**
2. **A swimming pool is half full of water. If the dimensions of the pool are 9 m by 12 m by 2 m, how many cubic metres of water are in the pool?**
3. **Find the volume of a cylinder that has a base radius of 10 cm and a height of 12 cm.**
4. **Julie has a waste bin that is cylindrical in shape. The base has a radius of 0.49 m and the height is 0.82 m. Find the volume of the bin.**
5. **A kart-racing track has a tunnel shaped like a triangular prism. The dimensions of the tunnel are: 13 m long, 3 m wide and 4 m from its highest point to the ground. Find the volume of the tunnel.**

Constructions

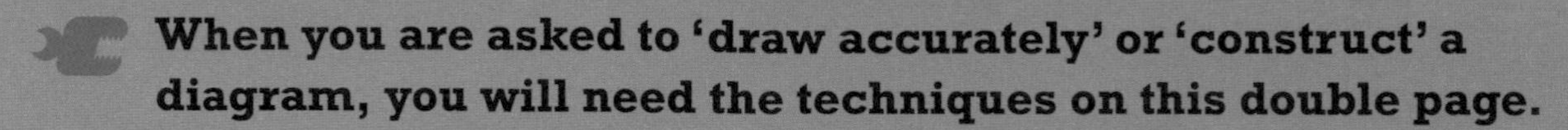

When you are asked to 'draw accurately' or 'construct' a diagram, you will need the techniques on this double page.

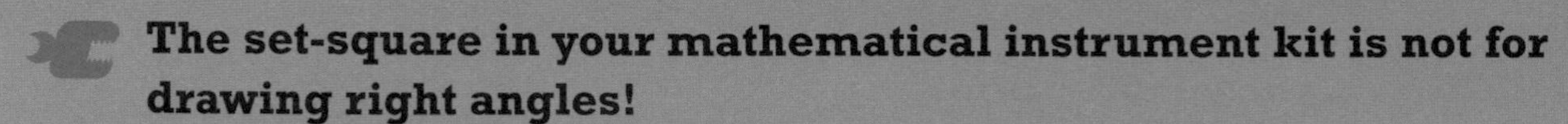

The set-square in your mathematical instrument kit is not for drawing right angles!

A Triangles

1. To draw a triangle accurately given all three sides: for example, 8 cm, 5 cm and 4 cm:

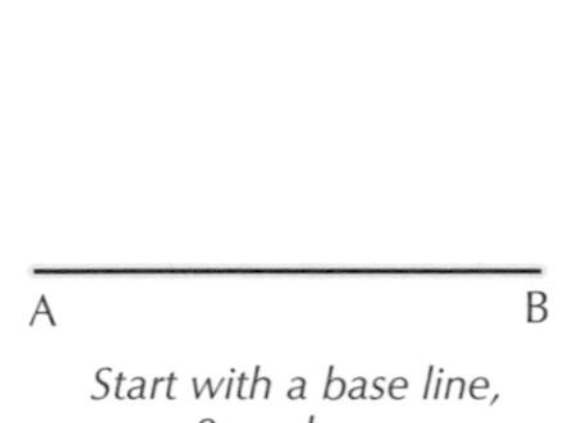

Start with a base line, 8 cm long.

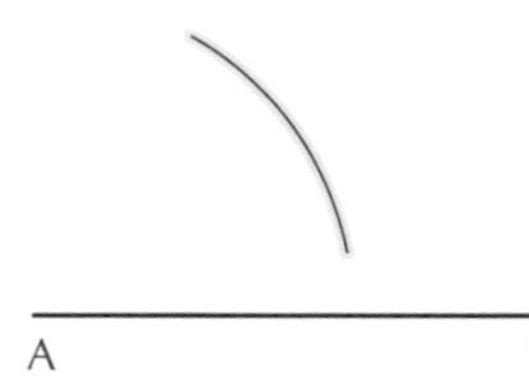

Set your compass to 5 cm. Put the point on A and draw an arc where you think the other vertex will be.

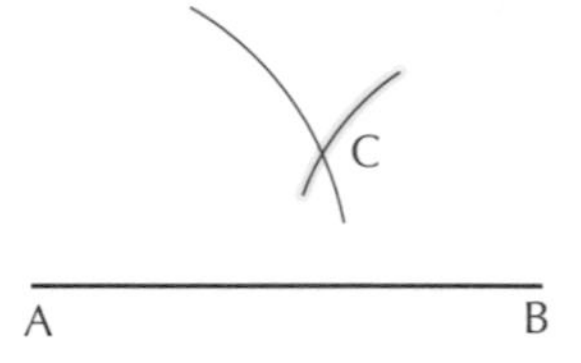

Set your compass to 4 cm. Put the point on B and draw another arc, crossing the first.

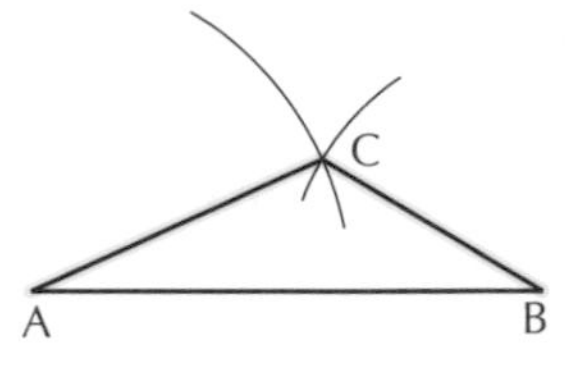

Join A and B to the point where the arcs cross.

2. To draw a triangle given one side and two angles: for example, 10 cm, with 50° at one end and 30° at the other:

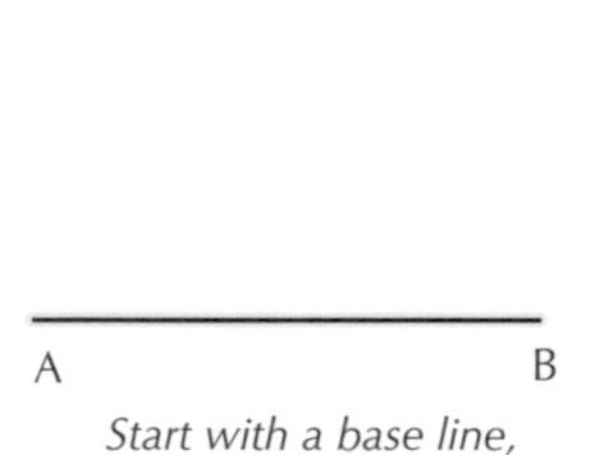

Start with a base line, 10 cm long.

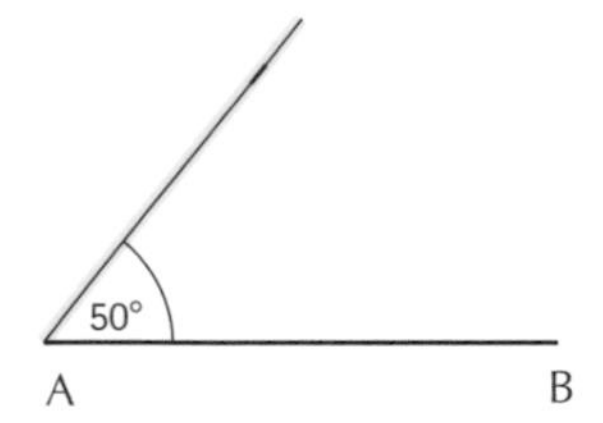

Draw a 50° angle at A. Make the line longer than you will need.

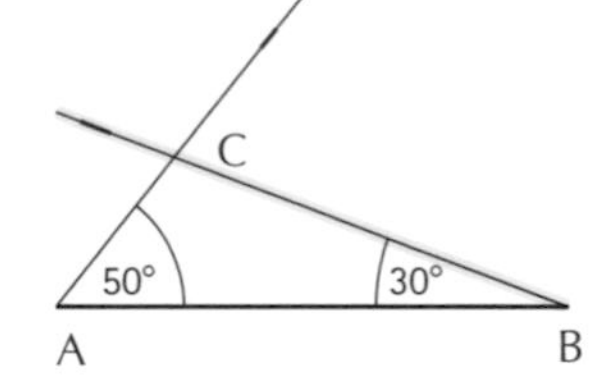

Draw a 30° angle at B. Make sure this line crosses the first.

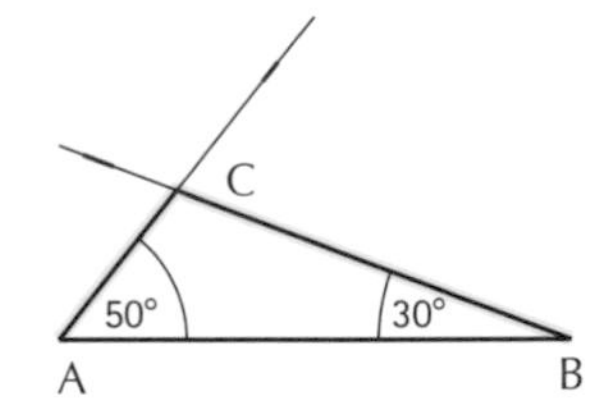

Join A and B to the point where the lines cross.

3. You can deal with other combinations of sides and angles by mixing the two methods.

>> **key fact** **To construct a 60° angle, just construct an equilateral triangle.**

B Perpendicular bisectors

1 **key fact** **A line that passes through the midpoint of two given points and is perpendicular to the line joining the points is called their perpendicular bisector.**

To construct a perpendicular bisector:

B
A

Mark the two starting points.

B
A

Set the compass to more than half the distance AB. Put the point on A. Draw arcs either side of the midpoint.

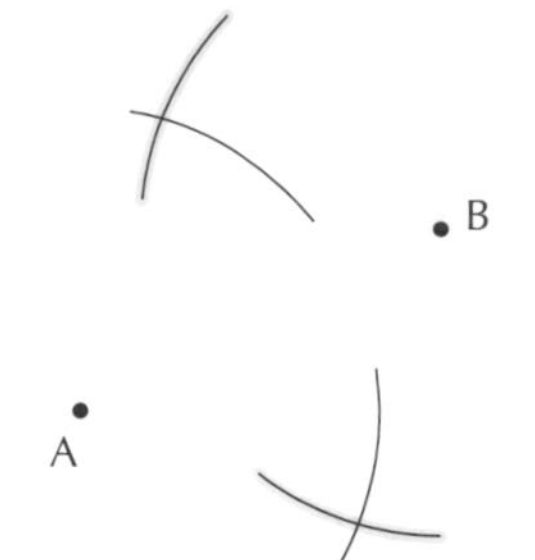

Keep the compass set to the same radius. Put the point on B. Cross the first set of arcs.

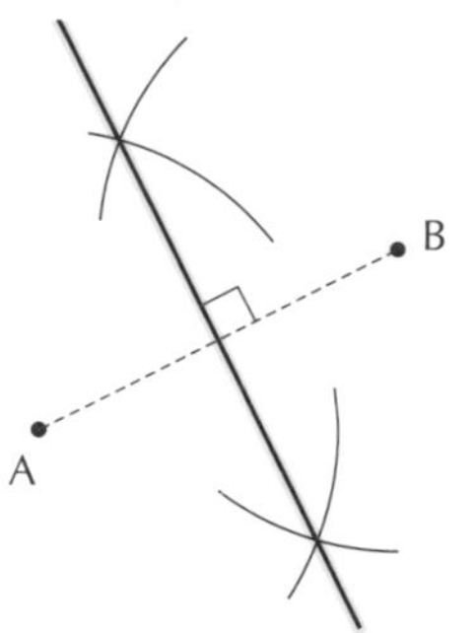

Joint the intersection points. This is the perpendicular bisector.

C Angle bisectors

1 **key fact** **To bisect an angle is to divide it exactly in half.**

2 Use this construction to bisect an angle.

Start with the angle you want to bisect.

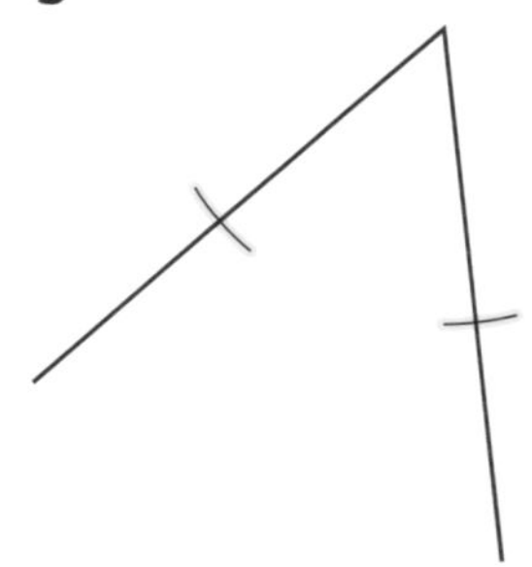

Set the compass to any radius. Put the point on the vertex. Draw arcs to cross both arms of the angle.

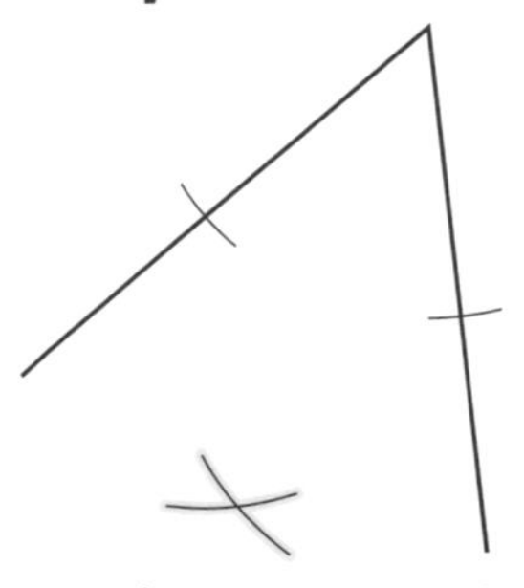

Keep the compass set to the same radius. Put the point on each of the intersections. Draw two arcs that cross as shown.

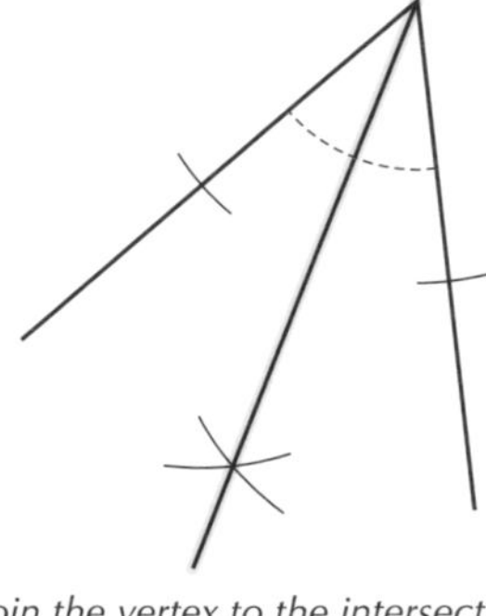

Join the vertex to the intersection. This line bisects the angle.

3 To construct a 45° angle, bisect a right angle: to construct 30°, bisect 60°.

>> practice questions

1 (a) Construct a triangle with sides of 10 cm, 8 cm and 6 cm.
(b) Construct the perpendicular bisectors of all three sides. They should cross at a single point.
(c) Put the point of your compass on the intersection and the pencil on one of the vertices. Draw a complete circle – it should pass through all three vertices.

2 (a) Construct a triangle with a 7 cm side, a 70° angle at one end and a 45° angle at the other.
(b) Bisect all three angles. The bisectors should cross at a single point.
(c) Put the point of your compass on the intersection. Adjust your compass and draw a circle that just touches all three sides of the triangle.

3 Using only your compass and ruler, construct angles of 60° and 90° next to each other to make an angle of 150°. Bisect this angle. Use your protractor to check that each half is 75°.

Loci

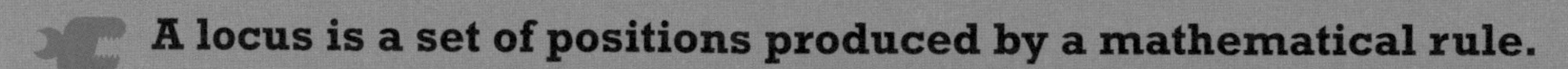
A locus is a set of positions produced by a mathematical rule.

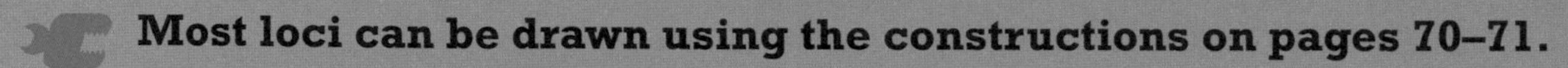
Most loci can be drawn using the constructions on pages 70–71.

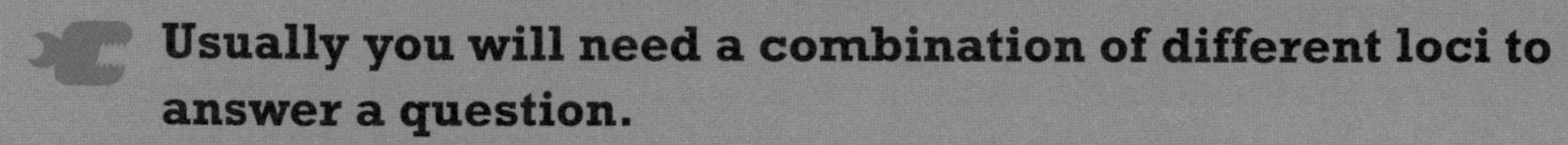
Usually you will need a combination of different loci to answer a question.

A What are loci?

1 **key fact** **A locus is a set of positions generated by some rule.**

Different loci are generated by different rules. A locus may be:

- part of a **line** or **curve**
- a **region** (area), inside or outside a shape.

2 Exam questions often depend on combining more than one rule to produce a locus.

B Fixed distances

1 The rule 'stay a fixed distance away from a fixed point' gives a circular locus.

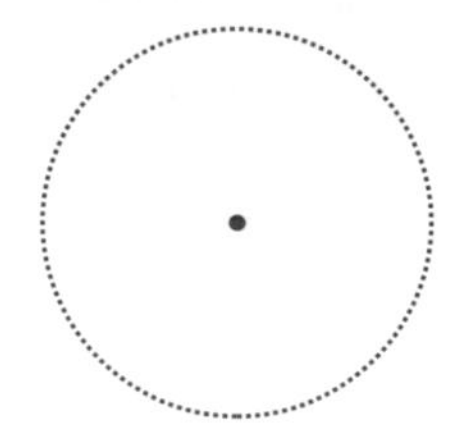

2 The locus of a point that stays a fixed distance from a straight line is two parallel straight lines.

3 You may need to combine these two locus types.

This locus is '2 cm from the line segment':

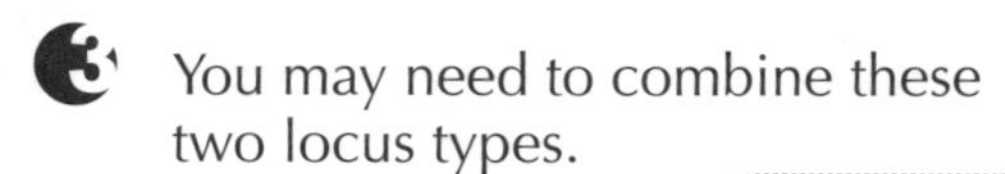

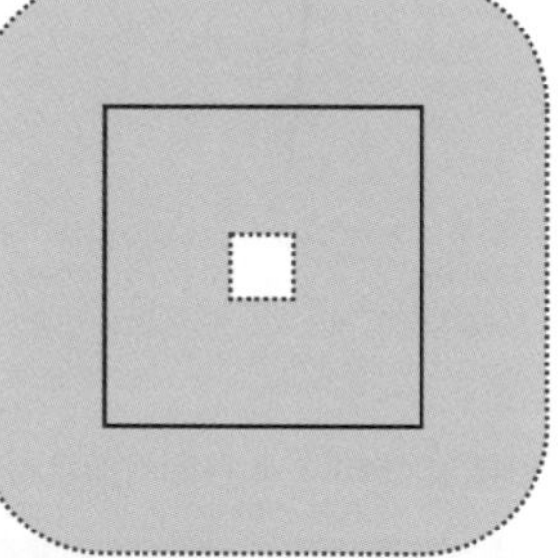

This locus is 'within 2 cm of the square'.

C Equal distances

1 **key fact** **A position that is the same distance from two objects is equidistant from them.**

The rule 'equidistant from two fixed points' is a straight line – the **perpendicular bisector** of the two points.

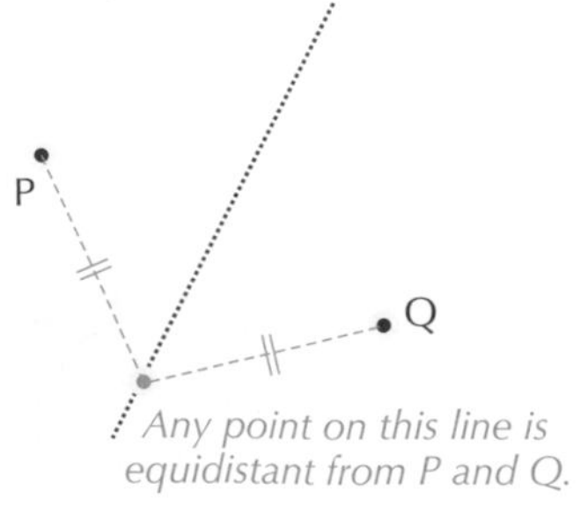

Any point on this line is equidistant from P and Q.

2 The rule 'equidistant from two straight lines' is a straight line – the **angle bisector** of the two lines.

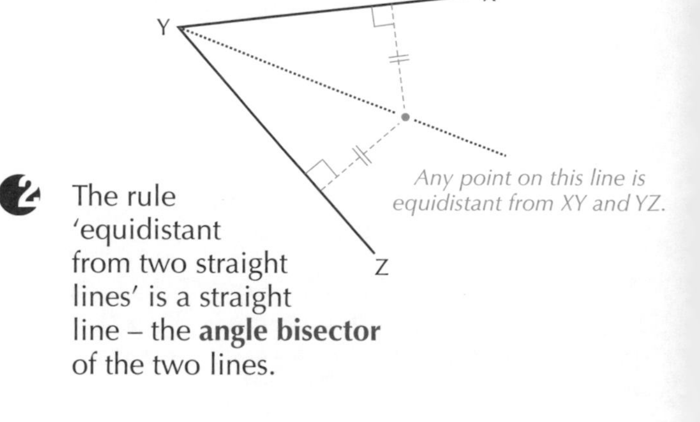

Any point on this line is equidistant from XY and YZ.

3 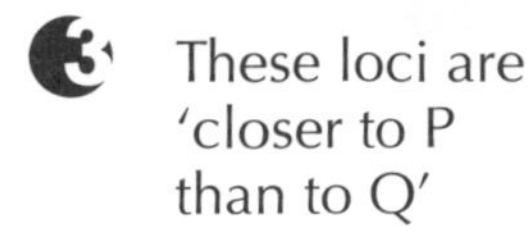
These loci are 'closer to P than to Q'

and 'closer to XY than to XZ'.

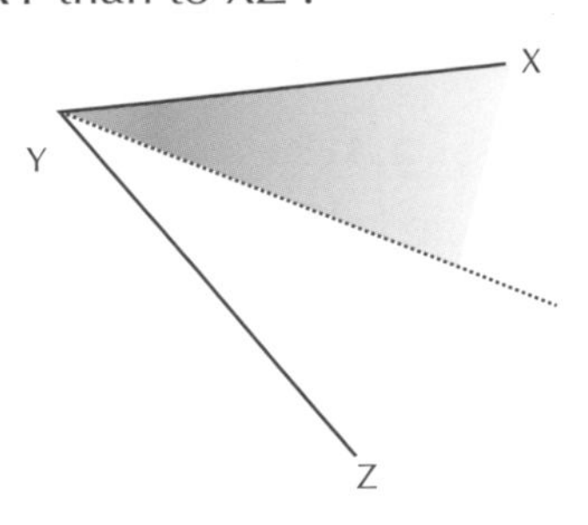

D Combining loci

1 Questions about combined loci usually involve:

- drawing two or more loci
- finding the points that are part of all the loci you have drawn.

2 The diagrams show how to answer the following question:

Point A is 4 cm from O. Indicate on your diagram the region containing points that are within 4 cm of O, and also closer to A than to O.

Mark O and A, 4 cm apart.

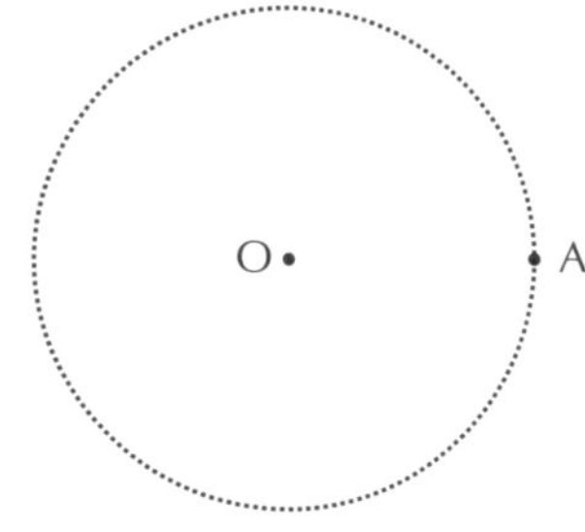

Set the compass to 4 cm. Draw a circle centred on O. It should pass through A.

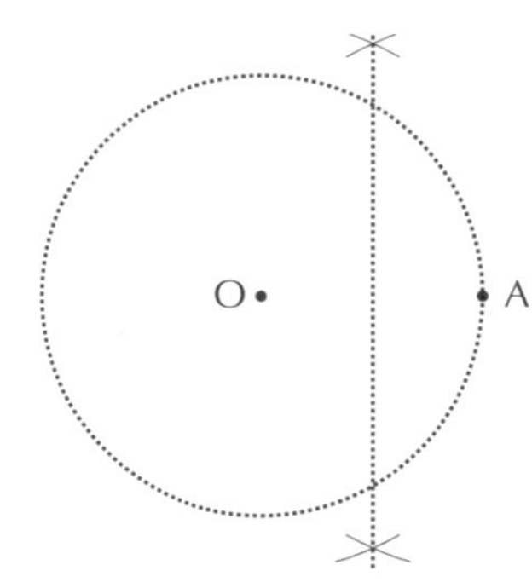

Use your compass to construct the perpendicular bisector for O and A.

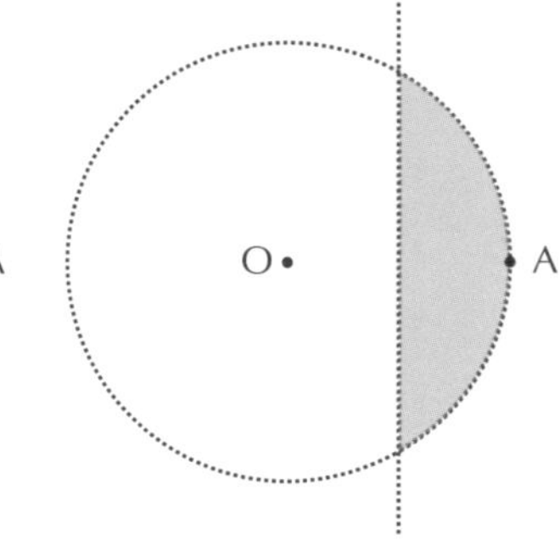

Shade the region inside the circle and to the right of the bisector.

>> practice questions

1 (a) Three radio transmitters R, S and T are positioned so that R is 90 km from S and 60 km from T; also, S is 75 km from T. Using a scale of 1 cm : 10 km, draw a diagram to show the positions of the transmitters.

(b) Broadcasts from R can be received up to 50 km away, from S up to 60 km away, and from T up to 35 km away. Indicate on your diagram the reception region for each broadcast. Shade in the region where all three broadcasts can be heard.

(c) How close could you be to S and still receive all three broadcasts?

2 (a) Draw an equilateral triangle with sides 7 cm long.

(b) Construct the locus of points that are exactly 1.5 cm from the triangle.

3 Draw a co-ordinate grid with x and y scales going from 0 to 10.

(a) Draw the locus of points that are equidistant from points (9, 2) and (9, 8).

What is the equation of this line?

(b) Draw the locus of points that are equidistant from points (2, 4) and (6, 8).

What is the equation of this line?

(c) Which point lies on both loci?

Transformations

- **A transformation changes shapes by altering their position or size.**
- **The original shape is called the object. A transformation produces the image shape.**
- **For enlargements, images are similar to their objects. For translations, reflections and rotations, the images are congruent to their objects.**

A Translation

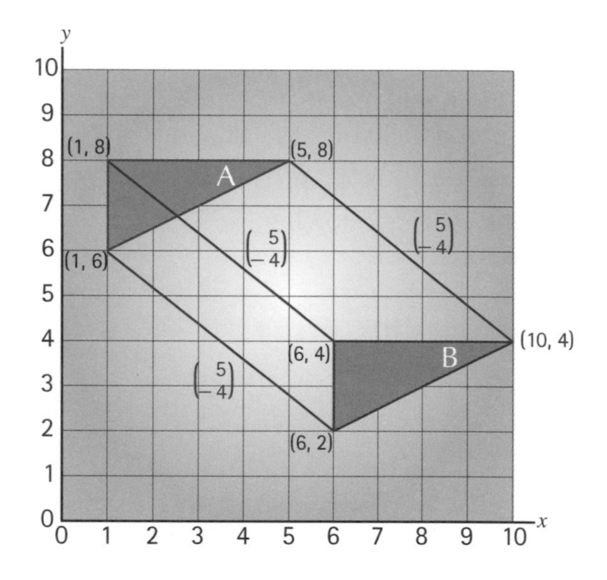

1 **key fact** **Translations are sliding movements. All they do is change the position of an object.**

Object triangle A in the diagram has been translated 5 units to the right and 4 units down to give image triangle B.

2 Translations can be described using **column vectors**. The vector for the translation shown is $\begin{pmatrix} 5 \\ -4 \end{pmatrix}$. Positive numbers are used for movements up or right, negative for down or left.

B Reflection

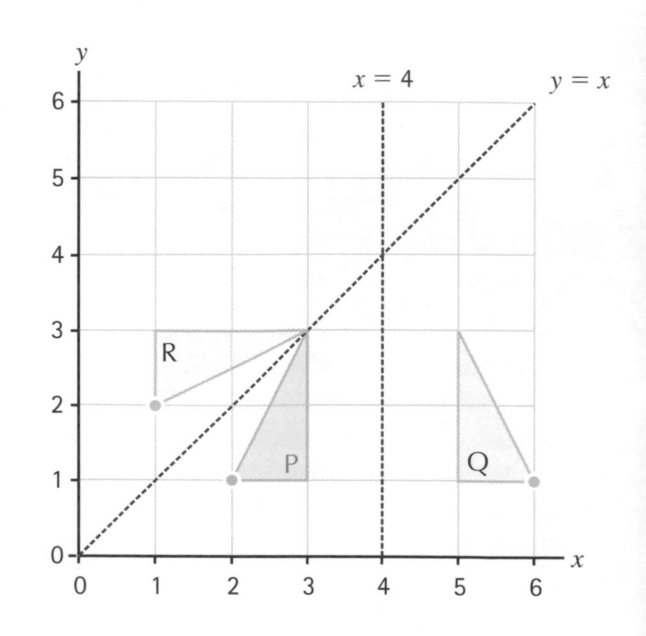

1 To describe a reflection, you need to specify where the mirror line is. The only mirror lines you will be asked to use are horizontal, vertical or at 45° to the axes.

>> **key fact** **Points in the object are the same distance from the mirror as their images.**

2 The diagram shows the object triangle, P, reflected in the line $x = 4$ to produce image Q. The point (2, 1) of P is 2 units away from the mirror; so is its image in Q, (6, 1).

3 In the diagram, P is also reflected in the line $y = x$ to produce image R. (2, 1) becomes (1, 2) in R: they are both half a *diagonal* unit away from the mirror.

C Rotation

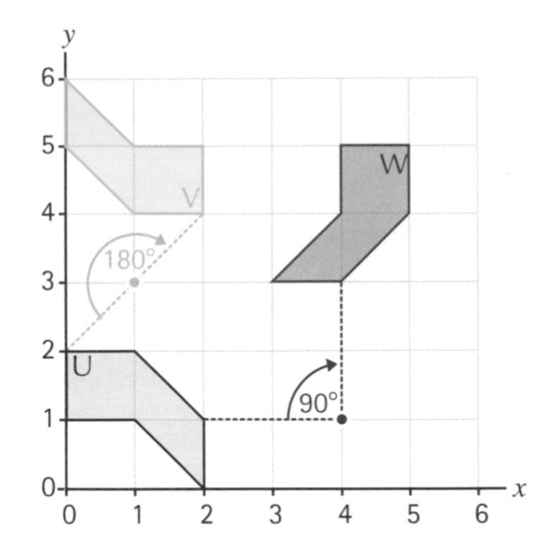

1 **key fact** **To describe a rotation, you need the angle the object turns through, and a fixed position called the centre of rotation.**

2 The only angles used in exam questions are 90° (clockwise or anticlockwise) and 180°. The diagram shows object shape U rotated by 180° about centre (1, 3) to give image V. It also shows U rotated 90° clockwise about centre (4, 1) to give image W.

D Enlargement

1 To describe an enlargement, you need the scale factor, and a fixed position called the **centre of enlargement**.

>> **key fact** **The scale factor controls how many times larger than the object the image is. Enlargements are mathematically similar to their objects.**

2 The diagram shows the object shape A enlarged using the origin as the centre. Image B was enlarged with scale factor 2, image C with scale factor 3.

3 Notice that you can reach the marked vertex of the object from the centre of enlargement using the vector $\begin{pmatrix}1\\2\end{pmatrix}$. If you keep using this vector, you find the corresponding points on the enlargements.

>> practice questions

For these questions, you will need to draw a co-ordinate grid with x- and y- axes from −10 to 10. Draw the object shape: a triangle with vertices (4, 0), (4, 2) and (3, 2).

1 For each part, write down the co-ordinates of the vertices of the image.

(a) Translate using the given vectors:

(i) $\begin{pmatrix}0\\6\end{pmatrix}$; **(ii)** $\begin{pmatrix}5\\3\end{pmatrix}$.

(b) Reflect in the following lines:

(i) $x = 5$; **(ii)** $y = x$.

(c) Rotate as follows:

(i) 90° clockwise, centre (−3, 4);

(ii) 180°, centre (2, −2).

(d) Enlarge with centre (6, 5):

(i) scale factor 2;

(ii) scale factor 3.

2 Describe fully the transformation that maps the vertices of the object to:

(a) (1, 0), (1, 2), (0, 2)

(b) (4, −6), (4, −8), (3, −8)

(c) (−1, 7), (−3, 7), (−3, 6)

(d) (−5, 0), (−5, 8), (−9, 8)

Statistical tables and diagrams

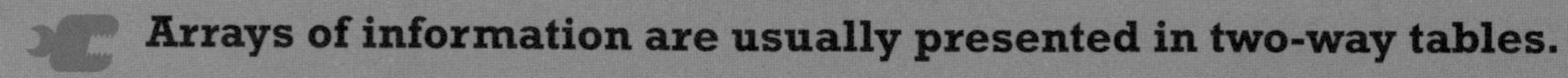

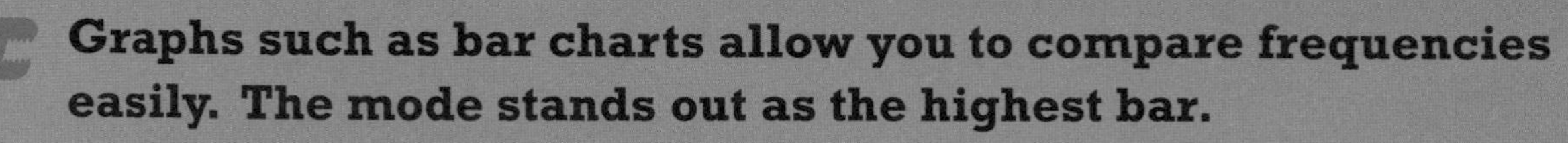

Arrays of information are usually presented in two-way tables.

Graphs such as bar charts allow you to compare frequencies easily. The mode stands out as the highest bar.

A Two-way tables

1. Two-way tables are used when information is arranged under two types of heading.

   This table shows the size of mathematics sets in a school. The cell highlighted in **red** shows that Set 3 in Year 9 contains 26 pupils.

	Set 1	Set 2	Set 3	Set 4	Set 5	Set 6	Set 7	Set 8
Year 7	31	30	27	27	27	22	20	12
Year 8	30	28	25	25	26	21	19	13
Year 9	31	30	**26**	26	26	22	21	15
Year 10	29	29	24	26	25	21	21	13
Year 11	30	29	26	25	25	21	20	10

   From this, you can find information such as how many pupils there are in any year group, or the whole school.

2. Bus and rail timetables are more examples of two-way tables.

B Stem-and-leaf diagrams

1. A stem-and-leaf diagram is a useful way of organising information.

   Suppose that these are the ages of people on a bus trip:

   22 16 45 20 32 41 28 22 31 25

   Put the information into ten-year groups:

```
1 | 6
2 | 2 0 8 2 5
3 | 2 1
4 | 5 1        Key: 1 | 0 = 10
```

>> **key fact** **The 'stems' are the numbers to the left of the vertical bar, representing the tens digit of each number. The 'leaves' are the units digits.**

   Then write each set of leaves in order:

```
1 | 6
2 | 0 2 2 5 8
3 | 1 2
4 | 1 5        Key: 1 | 0 = 10
```

   This is the finished diagram.

2. It is always easier to create stems and leaves first, then order the data.

C Frequency graphs

1 The information from the last section can be illustrated using one of these graphs.

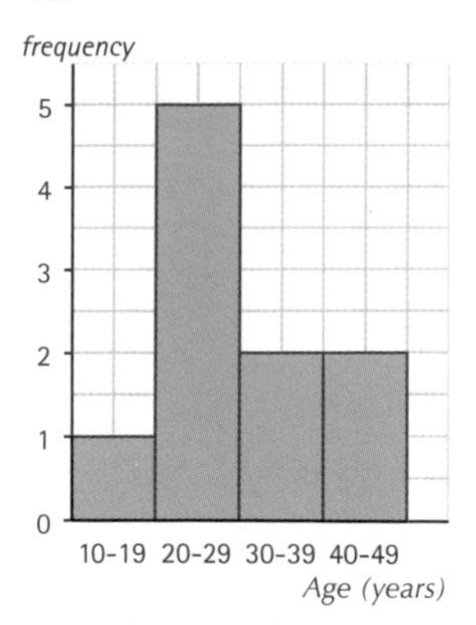

Bar chart or bar graph

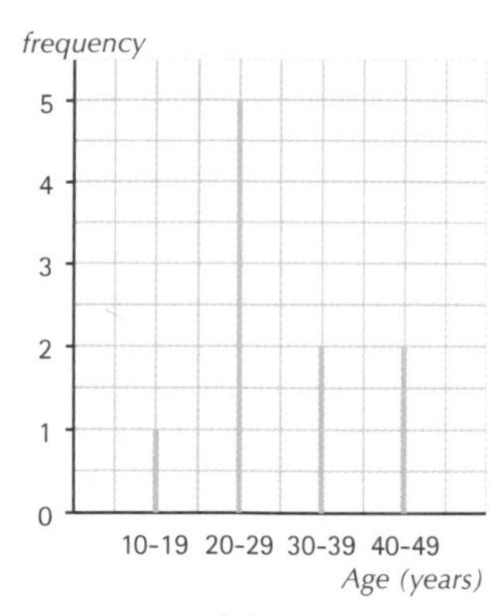

Vertical line graph

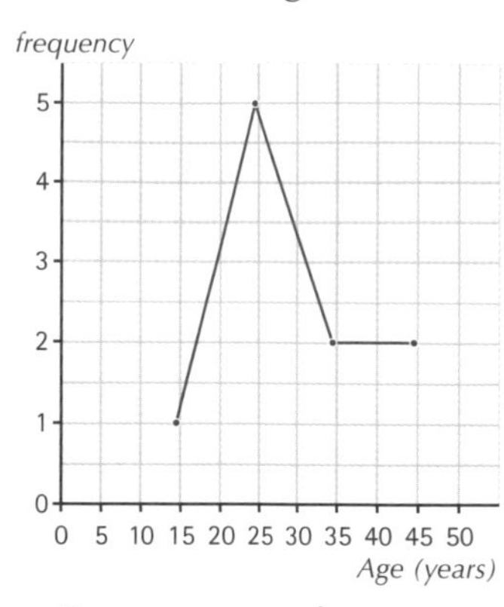

Frequency polygon

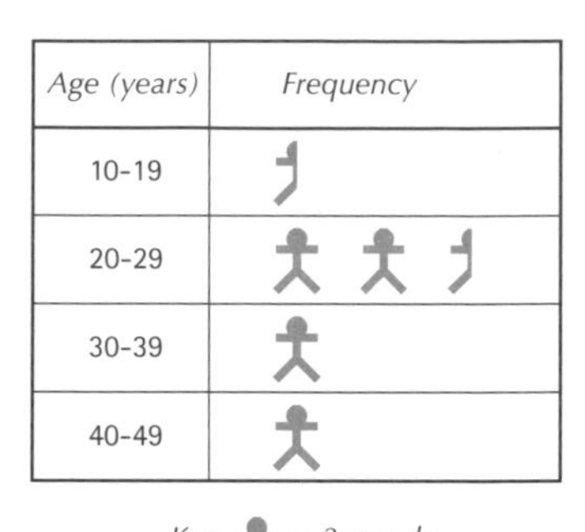

Pictogram

D Time-based graphs

1 You use a time-based graph to show how something changes with time. This graph shows how the value of a car falls over the years.

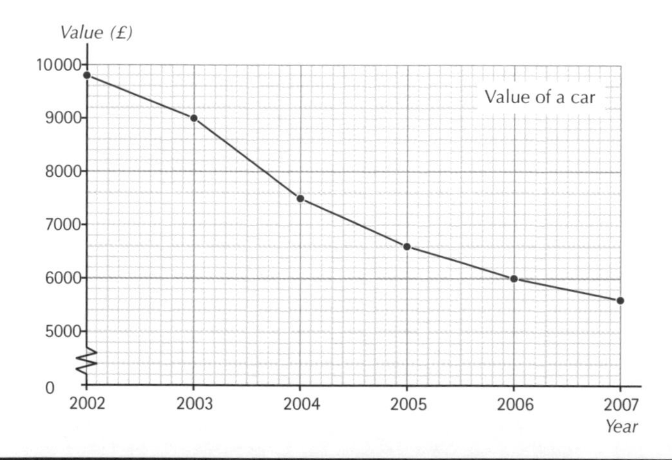

2 This graph shows a cycle ride. The steepness (gradient) of the graph tells you the speed of the cyclist.

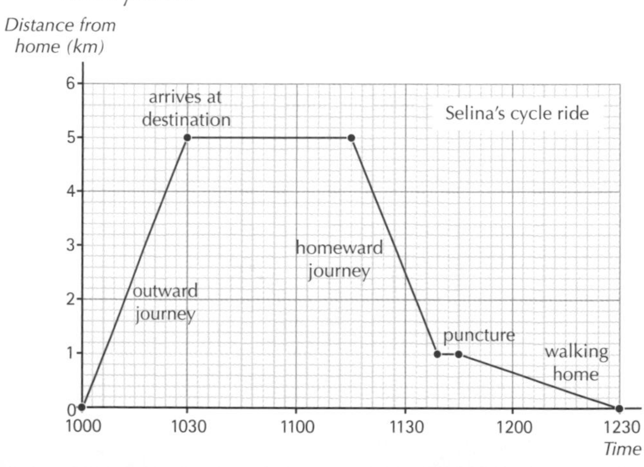

>> practice questions

1 These are the weights of a group of pupils, in kg.

55	52	70	65	72	47	48	60	53	45
29	51	50	72	47	50	48	51	29	45

(a) Construct a stem-and-leaf diagram of the data.

(b) Using the groupings from part (a), produce a bar chart of the data.

2 Use the graph in section D1 to answer these questions.

(a) How much was the car worth initially?

(b) In which year did the car's value fall the most? By how much did it fall?

(c) What percentage of its value did the car lose over the period shown?

3 Use the graph in section D2 to answer these questions.

(a) How long did Selina's outward journey take?

(b) What was her average speed for this?

(c) How long did she remain stationary when she had her puncture?

(d) What was her speed as she walked home with her bike?

Pie charts

Pie charts are useful when you want to see what fraction of a whole each item in a survey represents.

A What is a pie chart?

>> **key fact** **Pie charts are circular diagrams – the shape of a pie – split up to show sectors, or 'slices', that represent each part of the survey. By looking at the size of each sector, it is possible to estimate the fraction of the total for each data item.**

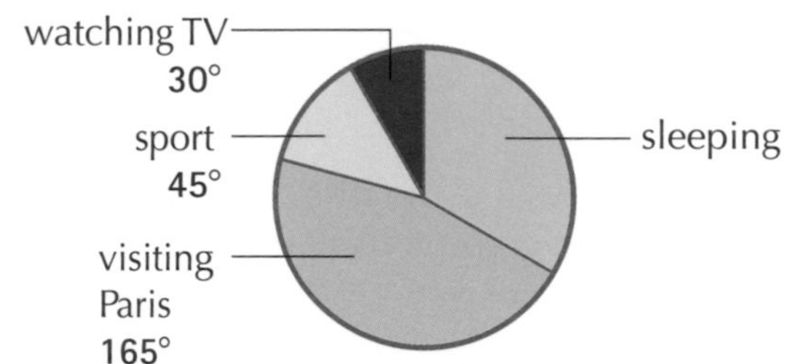

This pie chart shows how a group of students on a school trip in France spent the day and represents a full 24-hour day.

2 They had to travel a long way from their hostel to visit Paris.
What fraction of the day was spent sleeping?

To find the answer, you have to work out the angle for the sleeping sector.

A pie chart is a circle and there are 360° in a circle.

Add up the angles for the other sectors.

$30° + 45° + 165° = 240°$

So, the angle for the sleeping sector is:

$360° - 240° = 120°$

The fraction for sleeping is $\frac{120°}{360°}$, which cancels to $\frac{1}{3}$.

3 How many hours in the day were spent visiting Paris?

This is shown as 165° on the chart.

To find the answer you first have to work out what 165° is as a fraction of 360°.

$\frac{165°}{360°}$ cancels to $\frac{11}{24}$.

To find out how long this is in hours, multiply the fraction by 24 hours giving an answer of 11 hours.

B Calculating angles in a pie chart

1 This data was collected in a survey of 300 people's favourite holiday destinations:

60 Australia **40 UK** **150 Spain** **50 France**

You can see that Spain is the favourite place. It can also be shown on a pie chart. A pie chart has impact because it is **visual**.

>> **key fact** **To find the angle at the centre of the sector, calculate:** $\frac{\text{number of people}}{\text{total in survey}} \times 360°$

The results are:

Australia $\frac{60}{300} \times 360° = 72°$ UK $\frac{40}{300} \times 360° = 48°$

Spain $\frac{150}{300} \times 360° = 180°$ France $\frac{50}{300} \times 360° = 60°$

C Drawing the pie chart

1 To draw a pie chart, follow these steps:

- Draw a circle.
- Draw a straight line from the centre to the circumference – a radius (in an exam question, one or both of these steps might have been done for you).
- Measure out the angle for the first sector, starting from this line.
- Work round the pie chart one angle at a time.
- Label each sector with its title, and possibly its data value.

2 For the holiday data from section B, the result looks like this (note that there is no particular reason why the destinations should be shown in this order).

The angles have been marked on this diagram, but you would not normally do this.

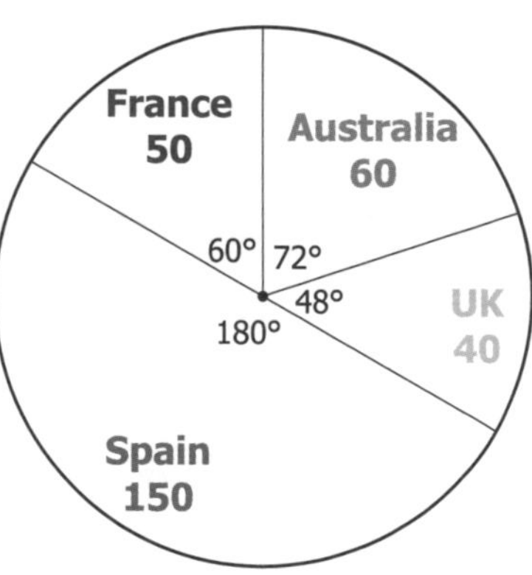

>> practice questions

1 This table contains data showing the number of films of each category shown in a city:

classification	18	15	12	PG	U	total
frequency	42	46	56	43	53	

Draw a pie chart to show this information.

2 Draw a pie chart to show the following data:

red balloons 30
yellow balloons 50
green balloons 120
white balloons 100

Finding averages

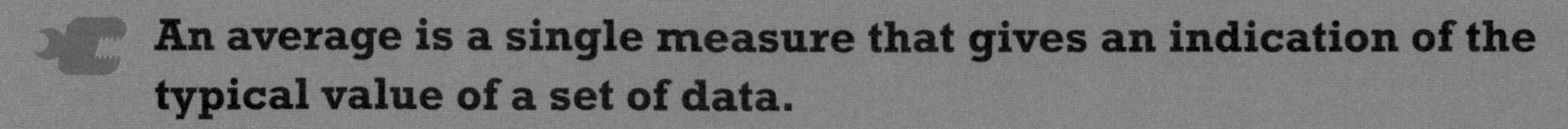

- An average is a single measure that gives an indication of the typical value of a set of data.
- There are three main types of average: mode, median and mean.

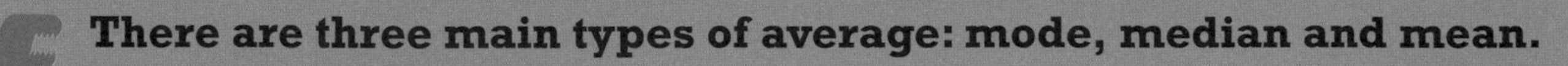

A The mode

>> **key fact** **The mode is the most common value in a set.**

1. This **frequency table** shows the number of pencils that children in one class have in their pencil cases:

number of pencils	0–4	5–9	10–14	15–19	20–24
frequency	8	5	4	12	2

The number of pencils has been placed in groups (e.g. 0–4), so there is some information we don't know. For example, we don't know how many children had eighteen pencils in their pencil case, even though it may be true that more children had eighteen pencils than any other number of pencils.

>> **key fact** **In the case of data in groups or classes, the group with the highest frequency is called the modal group.**

2. So, in this example it is not possible to give the mode as a single figure. What is clear is that more children had between 15 and 19 pencils than any other number of pencils. This is called the **modal group**.

B The median

>> **key fact** **The median is the middle value in a set, when all the numbers are arranged in order. If you have an even number of data items, the median is the mean of the middle two numbers.**

1. The median can also be found by totalling the frequencies.
2. In the table in A, above, the total is 31. Remember the median is the **middle term**, so count up from the left-hand side of the table.
3. The median must be the contents of the pencil case of the sixteenth child and since that child is in the 10–14 group, then the median lies in this group.

C The mean

>> **key fact** **The mean is the most frequently-used average. It is calculated by taking the sum of all the data items, then dividing the number of items.**

1. Suppose the weights of five newly hatched chicks are 7.2 g, 6.5 g, 10.0 g, 9.1 g and 5.7 g.

   The total weight of the chicks is $7.2 + 6.5 + 10.0 + 9.1 + 5.7 = 38.5$ g.

   The mean weight is therefore $38.5 \div 5 = 7.7$ g.

D The mean (grouped data)

1. In a frequency table, you multiply the data values by the frequencies.
2. In the example of the table opposite we do not know exactly how many pencils each child has. So it is not possible to calculate the mean accurately.
3. But if we assume that the items in each group are evenly spread, we can use the 'half-way value' to represent the group, and to find an approximate to the mean of this data.

number of pencils	f	mid-point, x	fx
0–4	8	2	16
5–9	5	7	35
10–14	4	12	48
15–19	12	17	204
20–24	2	22	44
total	**31**		**347**

4. Therefore, the mean number of items is approximately $347 \div 31 = 11.19$ pencils (to 2 decimal places).

>> practice questions

44 boxes of apples were examined and the number of damaged apples in each box was recorded:

damaged apples	0-4	5-9	10-14	15-19	20-24
frequency	15	10	9	6	4

1 **Estimate the mean value for this distribution.**

2 **What is the modal class for this distribution?**

3 **Find the median of this set of numbers: 2, 3, 1, 6, 2, 1, 3, 3, 4, 2.**

Comparing sets of data

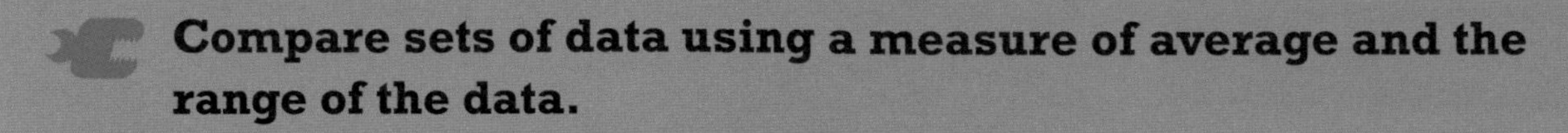
Compare sets of data using a measure of average and the range of the data.

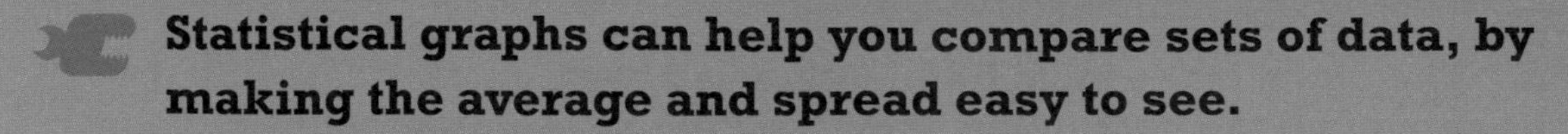
Statistical graphs can help you compare sets of data, by making the average and spread easy to see.

A Average and spread

1 **key fact** **You can compare two sets of data by:**

- **checking which is bigger on average (using mode, median or mean)**
- **checking which is more consistent (less spread out, using the range).**

2 This table shows the last five javelin throws by two athletes, in metres.

Mel	21	26	23	25	26
Kim	18	22	29	24	22

This table compares their results.

	furthest	shortest	mean	range
Mel	26	21	24.2	5
Kim	29	18	23	11

Kim threw the furthest, but also had the shortest throw.

Mel's mean throw was over a metre further than Kim's, and her range is smaller, so her performance was more consistent. If you were choosing a member for the javelin team, you might gamble on Kim, as she had the longest throw, but Mel is probably the better choice as she is more reliable.

B Back-to-back stem-and-leaf diagrams

1 This back-to-back stem-and-leaf diagram shows the lifetimes of two types of battery.

Duraplus		XtraPower
7	4	
4	5	4 6
8 6 2	6	0 3 8
9 8 8 6 5 4 4 1	7	2 2 3 5 7 7 8
7 7 5 5 2 2	8	0 0 1 3 3 3 6 9
8 4 3	9	5 6 6
3 2	10	

Key: 1 | 0 = 10 hours

2 Which are the better batteries? Duraplus have the longest lifetime, but their modal stem is only 70. XtraPower's modal stem is 80, so they last longer on average, and their data only covers 5 stems, as opposed to Duraplus's 7, so they are more reliable.

C Comparing graphs

1. This example uses frequency polygons to compare the time two groups of students spent revising for exams.
2. Student group A were more consistent in their revision, as their range is smaller. Both groups have a mode of 2 hours, but Group B contained more students who revised for longer than 2 hours.
3. There are many other comparisons you can make. For example, there was a student in Group B who did no revision, but every student in Group A did at least one hour, and so on.

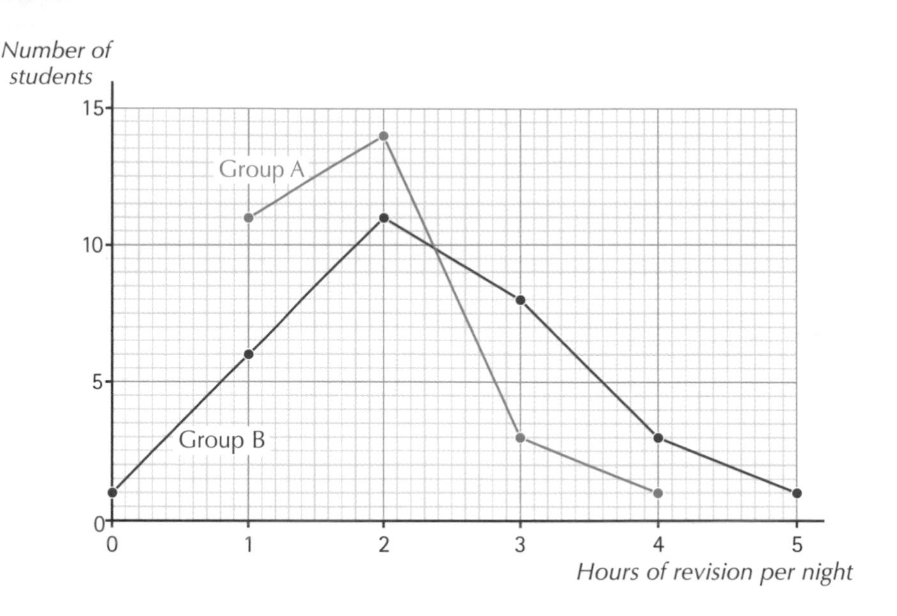

>> practice questions

1 Two types of marker pen are tested to see what length of line they write before running out of ink. Ten pens of each type were tested. These are the lengths, in metres, rounded to the nearest 10 m.

Markit	1710	1730	1650	1730	1680	1730	1680	1700	1670	1720
Skribbla	1640	1930	1870	1790	1800	1640	1720	1690	1720	1700

(a) Draw a table, like the one in section A, to compare the results.

(b) If you were going to buy a marker pen, which type would you buy, and why?

2 The same kind of plants were grown using two different fertilisers, Gromore and Sproutwell. The heights the plants grew to are shown in this bar chart, rounded to the nearest centimetre.

(a) How many plants were used in the experiment?

(b) Compare the results of the two fertilisers. Which fertiliser do you think was more effective?

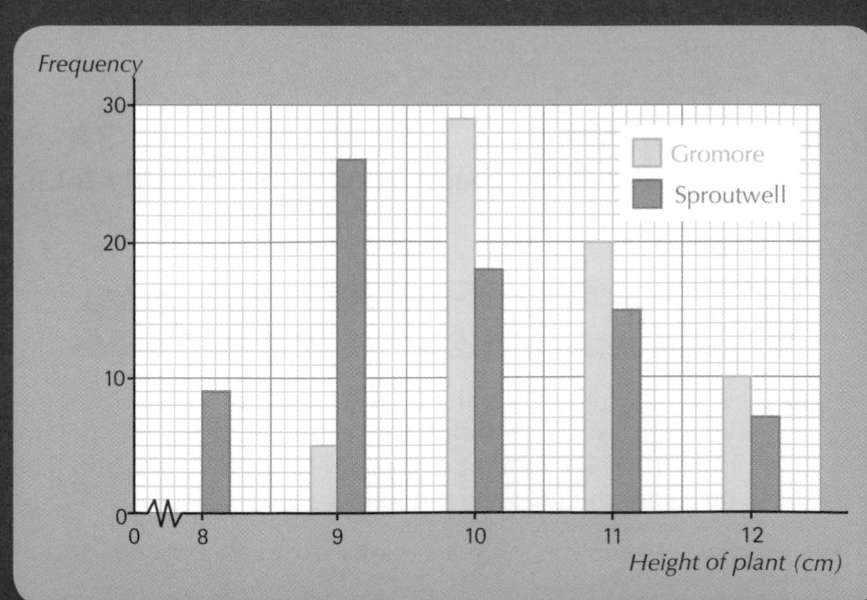

Probability

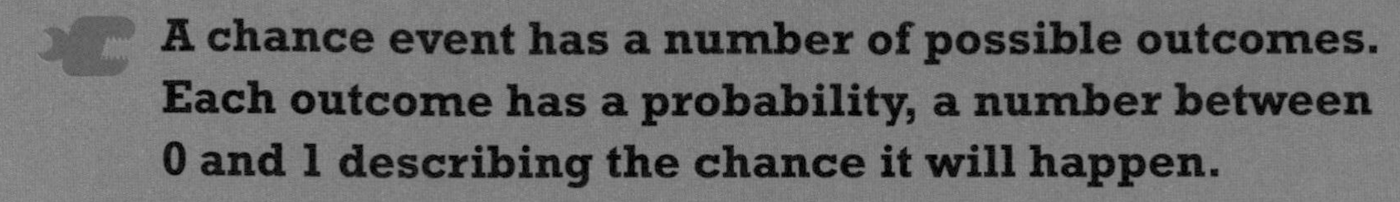

A chance event has a number of possible outcomes. Each outcome has a probability, a number between 0 and 1 describing the chance it will happen.

Probabilities can be expressed as fractions, decimals or percentages.

A Theoretical probability

1 The event 'roll a die' has 6 outcomes, all of which are equally likely – that is, they have the same probability, if the die is fair.
This is written

$P(1) = \frac{1}{6}$, $P(2) = \frac{1}{6}$, etc.

2 You can use the probability to estimate what will happen when a number of **trials** or experiments are carried out.

>> **key fact** **The expected frequency for an outcome is given by probability × number of trials.**

If you rolled the die 90 times, you would expect each number to come up roughly $\frac{1}{6} \times 90 = 15$ times. This is not a precise prediction, but the more trials you do, the better the prediction is likely to be.

3 You can also predict probabilities when the outcomes are not equally likely.
The probabilities for this spinner are

$P(\text{red}) = \frac{1}{4}$ and $P(\text{yellow}) = \frac{3}{4}$.

B Experimental probability

1 In some circumstances, it isn't possible to find probabilities just by analysing an event mathematically. You need to carry out a number of trials and use the results to estimate probabilities. This gives you the **experimental probability** (sometimes called **relative frequency**).

2 Suppose that in a survey of 500 cars, 35 had at least one defective tyre. You could estimate that the probability that a car checked at random has a defective tyre is $\frac{35}{500} = 0.07$.
The more data you have, the better the estimate of probability is likely to be.

>> **key fact** **Experimental probability for an outcome** $= \frac{\text{frequency of outcome}}{\text{number of trials}}$

C Possibility spaces

1 When two events with equally likely outcomes occur together, you can use a **possibility space table** to list all the possible outcomes. For example, if you roll two dice and add the scores, these are the possible outcomes:

		1st die					
		1	2	3	4	5	6
2nd die	1	2	3	4	5	6	7
	2	3	4	5	6	7	8
	3	4	5	6	7	8	9
	4	5	6	7	8	9	10
	5	6	7	8	9	10	11
	6	7	8	9	10	11	12

Each cell in the table is equally likely, so each has probability $\frac{1}{36}$.

This means that, as 7 occurs 6 times in the table, $P(7) = \frac{6}{36} = \frac{1}{6}$.

D Tree diagrams

1. Where two (or more) events occur together and the probabilities are *not* equal, a tree diagram is useful. Each event has a set of 'branches'. The outcomes are written at the end of each branch, with the probabilities along the branches.

2. This diagram illustrates the following situation: a motorist has to drive through two major sets of traffic lights on her way to work. At the first set, the probability that she will get through without stopping is 0.8; at the second, it is 0.7. With the diagram, you can find the probability that she will have to stop twice, once, or not at all.

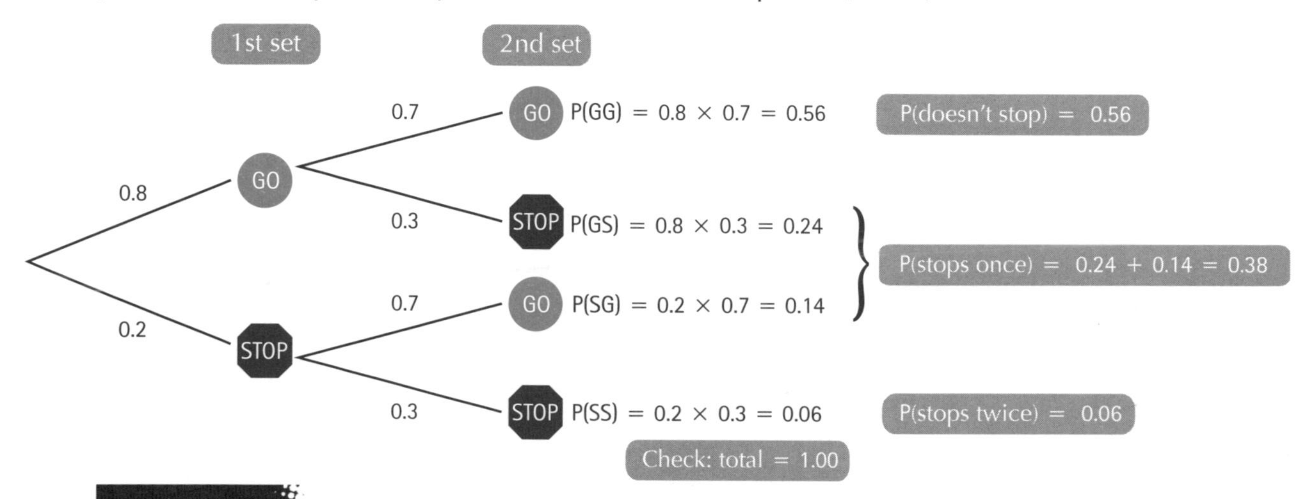

>> **key fact** **To obtain the correct probability for any route through the tree, multiply the probabilities on any branches you use. If the probabilities from two routes have to be combined, add them.**

>> practice questions

1 In a traffic survey, the number of people in each car was counted. These are the results.

Number of people	1	2	3	4	5
Frequency	144	75	36	33	12

Use this information to estimate the probability that a car picked at random:

(a) contains just the driver;

(b) contains at least three people.

2 In a charity game, you spin these two spinners. You win a prize if both spinners land on the same colour. Draw a tree diagram to find the probability of winning a prize.

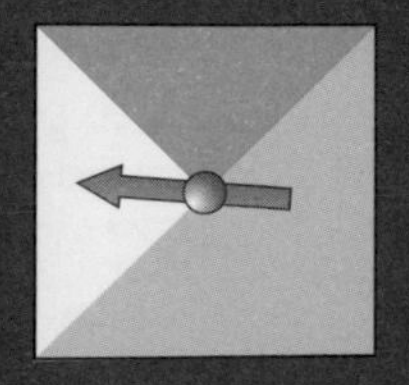

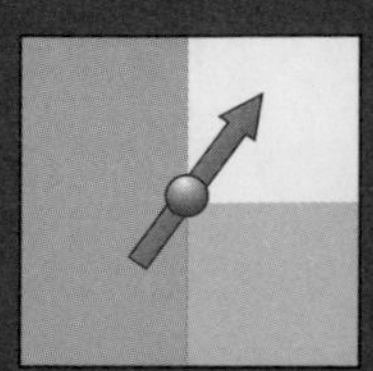

3 In a game, you roll two dice but score the difference between the numbers showing on the dice. For example, if you roll a six and a two, your score is 6 − 2 = 4.

(a) Draw a possibility space table to show all the possible scores.

(b) Which score is most likely, and what is its probability?

(c) If the dice are rolled 100 times, how many times would you expect to score 2?

Scatter diagrams and correlation

- **When you think there might be a link between two sets of data, draw a scatter diagram.**
- **If there is a link, the points will seem to be scattered close to a line, the line of best fit.**
- **The line of best fit can be used to make predictions about similar data.**

A Scatter diagrams

1. When it is thought that two sets of data may be linked, a scatter diagram can reveal this.

   For example, this table shows the scores obtained by 18 students in two maths tests.

test 1	5	10	15	20	25	30	35	40	45	50	55	60	75	80	85	90	95	100
test 2	4	1	5	16	28	37	46	39	37	57	60	61	86	90	82	81	97	100

2. The pairs of data values are plotted as co-ordinates.

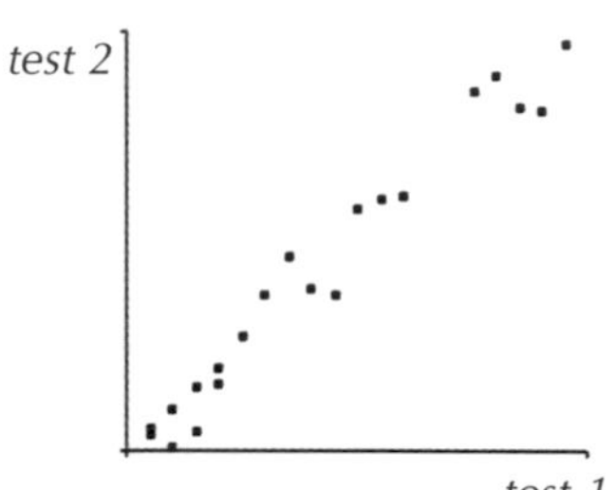

Each data point represents one person's scores. The correlation is positive.

>> **key fact** **You should never attempt to join scattered points to each other with straight lines!**

B The line of best fit

1. When your scatter diagram shows that there is a correlation, a **line of best fit** can be added to it.

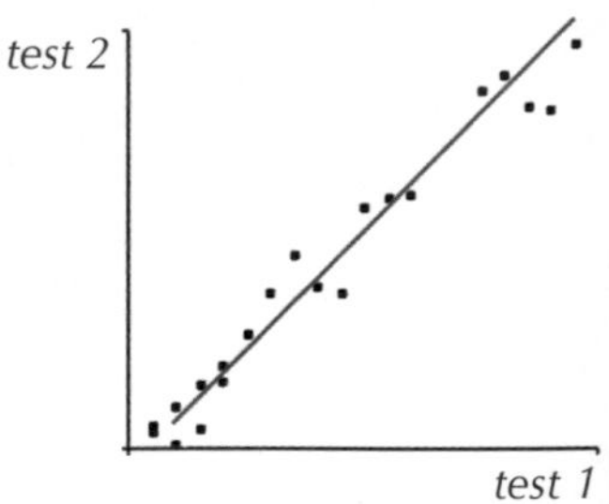

The line of best fit shows the 'trend' of the points. The points are close to the line, so the correlation is strong.

>> **key fact** **There should be roughly equal numbers of points either side of the line. Any points that lie a long way from the line are called outliers.**

2. You can use the line of best fit to predict values that follow the same trend. For example, you could answer the question, 'What would someone who scored 70 on test 1 expect to score on test 2?'

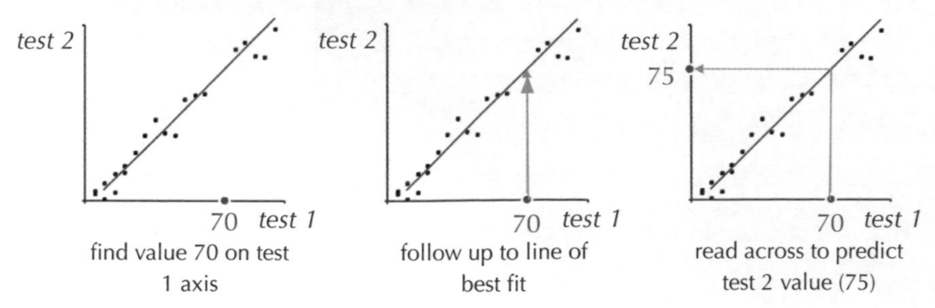

find value 70 on test 1 axis | follow up to line of best fit | read across to predict test 2 value (75)

3. Warning! Do not use the line of best fit to predict values outside the range of the data – the correlation may not hold for those values.

C Different types of correlation

>> **key fact** **Where the points are scattered closely around the line, there is a strong correlation.**

1 If the points are more loosely scattered around the line, there is a moderate correlation. This means that any predictions would be rough estimates, especially if the points are widely scattered, as they are in this diagram.

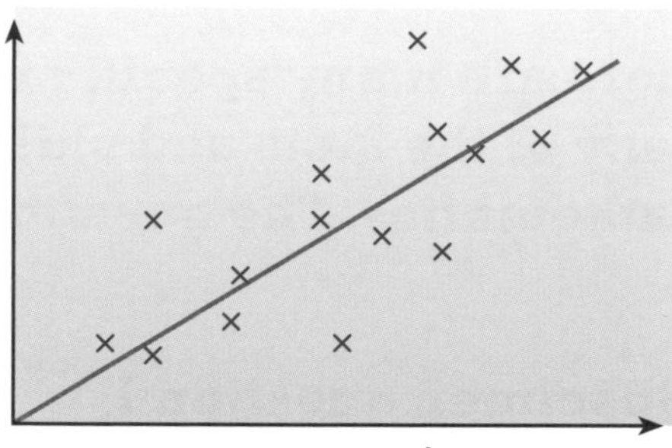

positive correlation

>> **key fact** **If the points are so scattered that there is no obvious line, then there is no correlation.**

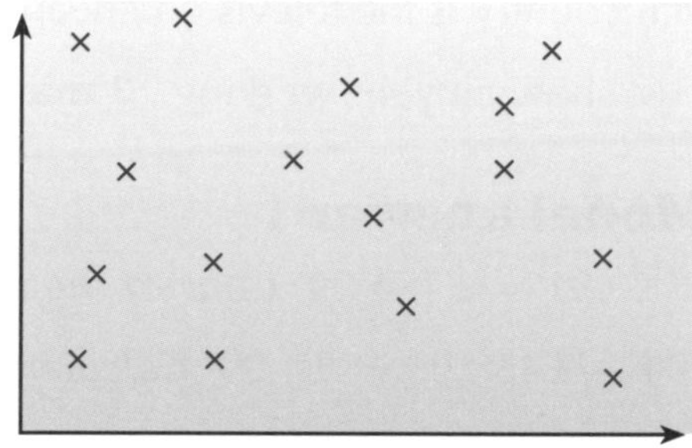

no correlation

2 If the points are scattered from top left to bottom right, the correlation is **negative**. This occurs when, as one quantity increases, the other quantity decreases.

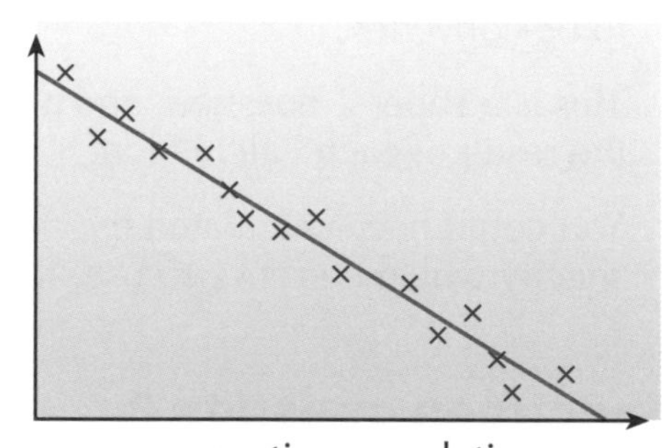

negative correlation

>> practice questions

This table shows the heights and weights of 10 people:

height (cm)	150	152	154	158	159	160	165	170	175	181
weight (kg)	57	64	66	66	62	67	75	69	72	76

1 **Draw the scatter diagram to represent the data. Add a line of best fit.**

2 **Comment on the correlation between height and weight.**

3 **Nisha weighs 58 kg. Estimate her height.**

(On one axis, use 2 cm to represent 5 cm of height and start at 145 cm. On the other axis, use 2 cm for 5 kg, starting at 55 kg.)

Exam questions and model answers

Number

There are many specific number questions on both papers, However, many of the facts and skills you learn are used in other branches of mathematics. The questions in this section test 'pure' number skills.

>> Specimen question 1

In Britain, a television costs £450. In Germany, the same television is sold for €499.
The exchange rate was €1 = 91 p on the day we checked.

In which country is the television cheaper, and by how much?

You must show all your working. **3 marks**

>> Model answer 1

499 × £0.91 = £454.09 ***Convert the German price in euros to pounds:* 1 mark**

£454.09 − £450.00 = £4.09 ***Calculate the difference between the two prices:* 1 mark**

The television is cheaper in Britain, by £4.09. ***Answer the question:* 1 mark**

- You can only compare the prices if they are in the same currency, so one of them has to be converted.
- This is a Paper 2 question, and although a calculator was used, you are asked to write the result of each calculation.
- You could have calculated the difference in euros, because the question doesn't specify which currency to use. To do so, convert £450 to 450 ÷ 0.91 = €494.51.

>> Specimen question 2

On a TV quiz show, teams buzz to answer a 'starter' question worth 10 points.

- If they interrupt the quizmaster and give a wrong answer, they lose 5 points.
- If they get the starter right, they also answer extra questions worth 5 points each.
- If they get a starter question wrong, but don't interrupt, they score no points.

These were the details of one programme.

	starters correct	starters wrong	starters wrongly interrupted	extra questions
New Bridge College	16	13	5	25
University of Lakeland	18	9	11	26

Who won the game? **3 marks**

>> Model answer 2

New Bridge College:

(16 × 10) + (13 × 0) − (5 × 5) + (25 × 5) = 160 + 0 + 25 + 125 = 260 points **1 mark**

University of Lakeland:

$(18 \times 10) + (9 \times 0) + (11 \times 5) + (26 \times 5) = 180 + 0 + 55 + 130 = 255$ points **1 mark**

New Bridge College won. ***Answer the question:* 1 mark**

- To do this question, calculate the total score for each team.
- This is a Paper 1 question, so the calculations need to be done mentally, or with extra working on the paper. The numbers have been kept easy to help you!
- An alternative (but unorthodox) method is to work out the difference between the teams for each column in the table. This would give Lakeland the following number of points more than New Bridge: $(2 \times 10) + (-4 \times 0) - (6 \times 5) + (1 \times 5) = 20 + 0 - 30 + 5 = -5$ (so Lakeland lost).

>> Further questions

1 (a) Use your calculator to find $\sqrt{7-1.53}$. Write down all the figures on your calculator.

(b) Write your answer to 4 significant figures.

2 Which of the following fractions is nearest to $\frac{3}{4}$? $\frac{11}{15}$ $\frac{23}{30}$ $\frac{34}{45}$ $\frac{43}{60}$

3 (a) An average human hair is 0.000 05 mm in diameter. Write this number in standard index form.

(b) One of the hairs from Michelle's head is 48 cm long. Assuming it is of average diameter, how many times longer is it than it is wide? Give your answer in standard index form.

4 Find the number exactly halfway between the given numbers.

(a) 0.41 and 0.44 (b) -2 and 12

5 Between which two consecutive whole numbers does $\sqrt{72}$ lie?

6 At a dinner, the ratio of vegetarians to non-vegetarians is 4 : 27. There are 216 non-vegetarians. How many vegetarians are there?

>> Answers to further questions

1 (a) 2.338 803 113... (b) 2.339

2 $\frac{3}{4} = \frac{135}{180} = 0.75$

$\frac{11}{15} = \frac{132}{180} = 0.7333...$

$\frac{23}{30} = \frac{138}{180} = 0.7666...$

$\frac{34}{45} = \frac{136}{180} = 0.7555...$

$\frac{43}{60} = \frac{129}{180} = 0.7166...$

So $\frac{34}{45}$ is closest.

3 (a) 5×10^{-5} mm

(b) $480 \div (5 \times 10^{-5}) = 9.6 \times 10^5$ times

4 (a) 0.425

(b) 5

5 8 and 9

6 $216 \div 27 = 8$;
$8 \times 4 = 32$ vegetarians

Algebra

Although algebra includes work on co-ordinates and graphs, the main skill you need is to be able to manipulate and transform algebraic expressions. With this skill, you can solve equations, transform formulae and analyse sequences.

>> Specimen question 1

Calculate the gradient of the straight line PQ. **2 marks**

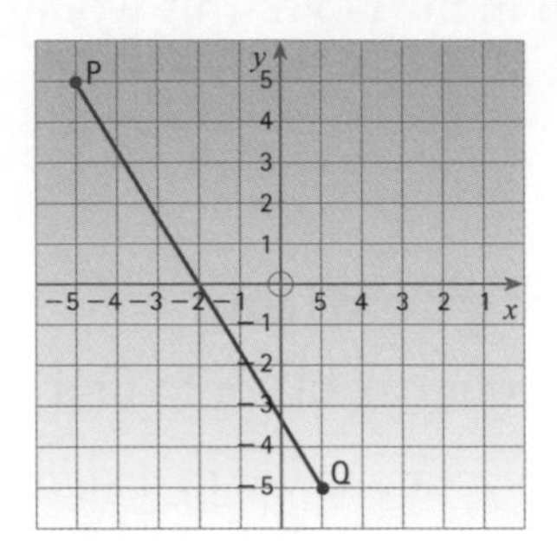

>> Model answer 1

Gradient $=$ increase in $y \div$ increase in x ***Write down the fact you are going to use***

$= -10 \div 6$ ***Substitute values:* 1 mark**

$= -1.67$ or $-1\frac{2}{3}$ ***Calculate the gradient:* 1 mark**

- Two main things can go wrong with a gradient calculation. One is getting the x and y increases mixed up. The other is not using a 'negative increase' when the graph slopes down from left to right.

>> Specimen question 2

Anthony weighed three bags of dried fruit, A, B and C. Bag B was 50 grams lighter than bag A. Bag C was three times as heavy as bag A.

(a) Using x to stand for the weight of bag A, write down expressions, in terms of x, for the weight of bag B and bag C. **2 marks**

(b) The total weight of the three bags was 840 grams. Write down and simplify the equation in terms of x. **2 marks**

(c) Solve your equation to find the weight of bag A. **2 marks**

>> Model answer 2

(a) Bag B: $x - 50$. ***Subtract 50 from x:* 1 mark**

Bag C: $3x$. ***Multiply x by 3:* 1 mark**

(b) $x + x - 50 + 3x = 840$ ***Add the expressions for bags A, B and C, making the result equal to 840:* 1 mark**

$5x - 50 = 840$ ***Simplify the left-hand side:* 1 mark**

(c) $5x - 50 = 840$

$5x = 840 + 50$ ***Add 50 to both sides***

$5x = 890$ ***Simplify***

$x = 890 \div 5$ ***Divide both sides by 5. At this stage you'll get the method marks: 1 mark***

$x = 178$ ***Simplify and state the answer: 1 mark***

- Remember always to do the same thing to both sides of your equation to keep it balanced.

>> Specimen question 3

(a) Expand and simplify $5(x + 1) - 2(x - 3)$. **2 marks**

(b) Use your answer to part (a) to solve the inequality $5(x + 1) - 2(x - 3) < 20$. **2 marks**

(c) Expand and simplify $(x + 1)(x - 4)$. **2 marks**

>> Model answer 3

(a) $5(x + 1) - 2(x - 3) = (5x + 5) - (2x - 6)$

$= 5x + 5 - 2x + 6$ ***Multiply each bracket out and subtract the second from the first: 1 mark***

$= 5x - 2x + 5 + 6$ ***Rearrange the terms if you like***

$= 3x + 11$ ***Simplify and state the answer: 1 mark***

(b) $3x + 11 < 20$ ***Use your expression from part (a)***

$3x < 20 - 11$ ***Subtract 11 from both sides***

$3x < 9$ ***Simplify***

$x < 9 \div 3$ ***Divide by 3***

$x < 3$ ***Simplify: 2 marks***

(c) $(x + 1)(x - 4) = x(x - 4) + 1(x - 4)$ ***Multiply each bracket out***

$= x^2 - 4x + x - 4$ ***Write down all four terms: 1 mark***

$= x^2 - 3x - 4$ ***Simplify: 1 mark***

- Be careful when subtracting a bracket containing a minus sign. The term being subtracted inside the bracket will change the sign to positive.

>> Specimen question 4

A sequence of numbers begins 2, 7, 12, 17, 22, …

(a) Find an expression for the nth term of the sequence. **2 marks**

(b) Calculate the 500th term of the sequence. **1 mark**

>> Model answer 4

(a) Each number in the sequence is 5 more than the last.

This means that the nth term must contain $5n$. ***Find the number multiplying n:* 1 mark**

If the nth term were $5n$, the first term would be $5 \times 1 = 5$, the second would be 10, and so on.

These terms are always three less than 5, 10, 15, …

So the nth term is $5n - 3$. ***Find the number added or subtracted from 5n:* 1 mark**

- Remember that if the terms count up in 5s, it doesn't mean that the nth-term expression is $n + 5$! This gives the sequence 6, 7, 8, 9, …

(b) When $n = 500$, the nth term $= 5n - 3$ ***Write down the formula***

$= 5 \times 500 - 3$ ***Substitute for n***

$= 2500 - 3$ ***Simplify***

$= 2497$ ***Calculate the value:* 1 mark**

>> Further questions

1 (a) Expand and simplify $4(3x + 5) - 2(x - 1)$ (b) Expand and simplify $(3x + 5)(x - 1)$

2 Solve the equations:
(a) $15 - 4x = 9$ (b) $5(3x - 2) = 20$

3 Make d the subject of the formula $S = 4 - 3d$

4 Solve the equation $x^3 - 4x = 10$ by trial and improvement, correct to 1 decimal place.

>> Answers to further questions

1 (a) $10x + 22$ (b) $3x^2 + 2x - 5$

2 (a) $x = 1.5$ (b) $x = 2$

3 $d = \frac{(4 - S)}{3}$

4

x	$x^3 - 4x$	comments
1	−3	$x > 1$
2	0	$x > 2$
3	15	$x < 3$
2.5	5.625	$x > 2.5$
2.6	7.176	$x > 2.6$
2.7	8.883	$x > 2.7$
2.8	10.752	$x < 2.8$
2.75	9.796875	$x > 2.75$

$x = 2.8$ to 1 dp

Shape, space and measures

This part of mathematics covers many different topics, including all work on angles, areas and volumes of shapes, and anything to do with measurement, including speed.

>> Specimen question 1

(a) Calculate the circumference of a circle of diameter 41 m. State the units of your answer. **2 marks**

(b) Calculate the area of a circle of radius 7 mm. State the units of your answer. **3 marks**

>> Model answer 1

(a) Circumference $= \pi \times$ diameter ***Write down the circle formula***

$= \pi \times 41$ ***Substitute the value for the diameter: 1 mark***

$= 128.8052...$ m ***Calculate – write down plenty of digits***

$= 128.8$ m to 1 dp ***Write the answer, rounded to a sensible accuracy: 1 mark***

(b) Area $= \pi \times \text{radius}^2$ ***Write down the circle formula***

$= \pi \times 49$ ***Substitute the value for the radius: 1 mark***

$= 153.9380...$ mm^2 ***Calculate – write down plenty of digits: 1 mark***

$= 154$ mm^2 to the nearest square mm ***Write the answer, rounded to a sensible accuracy: 1 mark***

- You are expected to know these formulae for the circumference and area of a circle – they are not on the page of formulae.
- Your calculator probably has a π key. Use this to get the greatest accuracy. If you don't have a π key, you will find a value to use (usually 3.142) on the front cover of the paper, or on the page of formulae.
- Use a rough estimate to check the answers to calculations like this: so in **(a)**, use $\pi \approx 3$ and diameter ≈ 40 to get the estimate $3 \times 40 = 120$ m. This is quite close to the calculated answer.

>> Specimen question 2

The diagram shows the side of a gardener's cold frame.

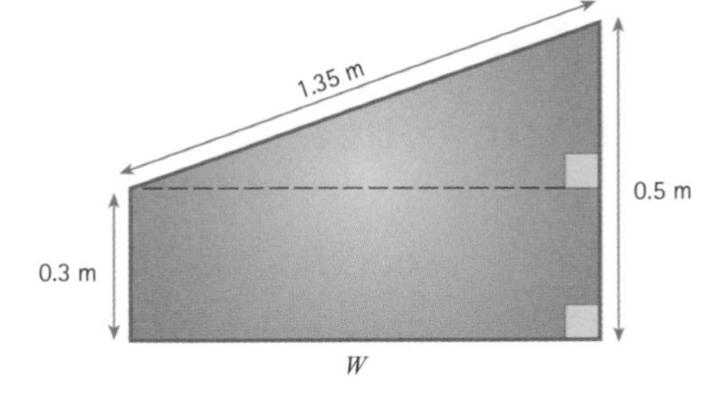

(a) Calculate the width (W in the diagram) of the frame. Give your answer to an appropriate degree of accuracy. **4 marks**

(b) Calculate the area of the end of the cold frame. **3 marks**

(c) The cold frame is a prism 1.5 m long. Calculate the volume of the frame. **2 marks**

>> Model answer 2

(a) In the triangle at the top of the diagram, the base is equal to W and the height is $0.5 - 0.3 = 0.2$ m. ***You either write this down, draw a diagram like the one below, or mark these lengths on the original diagram: 1 mark***

Using Pythagoras' Rule, $W^2 = 1.35^2 - 0.2^2$ ***State Pythagoras' Rule and substitute values: 1 mark***

$W^2 = 1.7825$

$W = \sqrt{1.7825}$ ***Calculate the square root: 1 mark***

$W = 1.335102...$ ***Keep this answer in your calculator for the next part of the question***

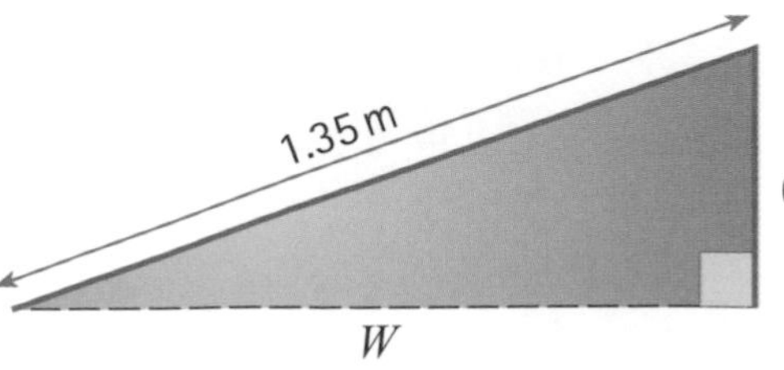

The width of the frame is 1.335 m, to the nearest millimetre. ***Give the rounded answer: 1 mark***

- It is very important to identify the height of the triangle, as you can't get any further without it.
- Remember that you are not finding the hypotenuse in this triangle, so you subtract values in Pythagoras' Rule.
- A rounding to the nearest centimetre (1.34 m) would be acceptable. However, nothing less accurate than this will do, because one of the given measurements is already given to the nearest centimetre.

(b) The shape is a trapezium. The formula for this is given at the front of the paper.

$a = 0.3$ m, $b = 0.5$ m, $h = W = 1.335...$ ***Write down the values you're going to use in the formula. It doesn't matter which is a and which is b***

Area $= \dfrac{(a+b)h}{2} = \dfrac{(0.3+0.5)\times 1.335...}{2}$ ***Write down the formula and substitute values: 1 mark***
Use the value for h from your calculator, not the rounded answer from part (a)

$= 0.534041...$ ***Carry out the calculation: 1 mark***
Keep this value in your calculator for the next part

$= 0.534$ m^2, to 3 sf. ***Write down a suitable rounded answer: 1 mark***

(c) Volume of a prism = area of cross-section × length

$= 0.534... \times 1.5$ ***Write down the formula and substitute values: 1 mark***
Use the area value from your calculator, not the rounded answer from part (b)

$= 0.801061...$

$= 0.801$ m^3, to 3 sf. ***Carry out the calculation and write down a suitable rounded answer: 1 mark***

- Remember to keep each answer in your calculator to use in the next part. You might introduce a rounding error otherwise, and lose marks.

>> Further questions

1 (a) Draw a co-ordinate grid with x-and y-axes from -2 to 12.

On your grid, draw a trapezium with vertices A(1, 2), B(1, 3), C(3, 5) and D(3, 1).

Calculate the area of trapezium ABCD.

(b) $A_2(5, 2)$ and $D_2(11, -1)$ are part of an enlargement of ABCD.
Plot A_2 and D_2 on your grid. What is the scale factor of the enlargement?

(c) Draw the rest of the enlargement. Write down the co-ordinates of B_2 and C_2.

(d) What are the co-ordinates of the centre of enlargement?

(e) What is the area of the enlargement?

2 The diagram shows the dimensions of a pentagonal prism.

(a) Sketch the net of the prism. Mark on relevant measurements.

(b) Calculate the surface area of the prism.

(c) Calculate the volume of the prism.
Give your answer in cubic metres.

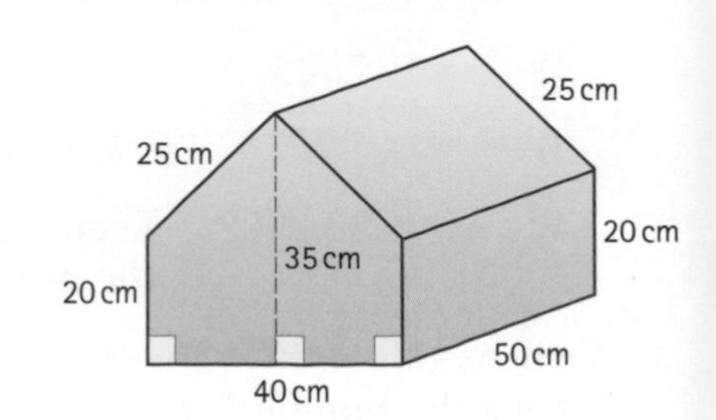

>> Answers to further questions

1 (a) 5 square units

(b) 3

(c) $B_2(5, 5)$ and $C_2(11, 11)$

(d) $(-1, 2)$

(e) 45 square units

2 (a)

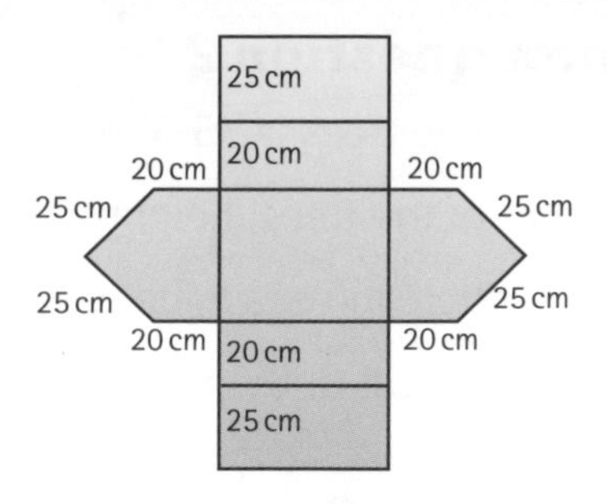

(b) One end $= (40 \times 20) + \frac{1}{2}(40 \times 15)$
$= 1100\,\text{cm}^2$
Rectangles $= 130 \times 50 = 6500\ \text{cm}^2$
Total $= 8700\ \text{cm}^2$

(c) $1100 \times 50 = 55\,000\,\text{cm}^3 = 0.055\,\text{m}^3$

Handling data

Handling data covers three main areas: representing data, which involves displaying data in charts and tables, and interpreting it; processing data, which includes work on averages and ranges to describe and compare frequency distributions; and probability, the study of chance events.

>> Specimen question 1

Here are the first 50 digits of π: 3.141 592 653 589 793 238 462 338 327 950 288 411 971 693 993 751 0
This is a frequency table to show how many times each digit occurs. For example, the digit 6 occurs three times in the first 50 digits.

Digit	0	1	2	3	4	5	6	7	8	9
Frequency	2	6	5	9	3	5	3	4	5	8

A digit is chosen at random. What is the probability the digit will be prime? **3 marks**

>> Model answer 1

The prime digits are 2, 3, 5 and 7. ***Show the examiners you know what the prime numbers are by writing them down or marking them on the table: 1 mark***

The total frequency of the primes $= 5 + 9 + 5 + 4 = 23$. ***Add the frequencies: 1 mark***

The probability is $\frac{23}{50}$. ***Write the fraction: 1 mark***

- This is a Paper 1 question. It is fine to leave the fraction just as it is, although you would be expected to cancel down if possible.

>> Specimen question 2

The mean height of five trees is 35.6 m.

(a) Find the total of the trees' heights. **2 marks**

One of the trees is rotten and has to be cut down. This changes the mean height of the remaining trees to 39.1 m.

(b) Calculate the height of the tree that was cut down. **2 marks**

>> Model answer 2

(a) $35.6 \times 5 = 178$ m. *Mean = total ÷ 5, so reverse the calculation:* **2 marks**
1 mark is for assembling the correct calculation

(b) New total $= 39.1 \times 4 = 156.4$ m. *Similar calculation to part (a):* **1 mark**

$178 - 156.4 = 21.6$ m *The difference between the two totals gives the answer:* **1 mark**

- A wrong answer from part **(a)** would be followed through.

>> Specimen question 3

A new exam is supposed to last two hours. In a trial of the new exam, a group of pupils were asked to try to finish the paper in two hours, but were given extra time if they needed it. This table shows the results.

Time taken (t minutes)	Frequency		
$60 \leq t < 90$	3		
$90 \leq t < 100$	8		
$100 \leq t < 110$	15		
$110 \leq t < 120$	42		
$120 \leq t < 130$	83		
$130 \leq t < 150$	49		

No student took less than 1 hour or longer than $2\frac{1}{2}$ hours.

(a) Calculate an estimate of the mean time taken, correct to the nearest minute. **5 marks**

(b) What is the probability that a student selected at random from this group took two hours or less to finish the exam? **2 marks**

(c) Comment on whether you think the exam contained too few questions, too many, or was about right. Explain your answer. **1 mark**

>> Model answer 3

(a)

Time taken (t minutes)	Frequency	Midpoint of group	Midpoint × Frequency
$60 \leq t < 90$	3	75	225
$90 \leq t < 100$	8	95	760
$100 \leq t < 110$	15	105	1575
$110 \leq t < 120$	42	115	4830
$120 \leq t < 130$	83	125	10 375
$130 \leq t < 150$	49	140	6860
	200		**24 625**

The data are grouped, so you have to make an estimate of the total for each group using the midpoints — these are the means of the highest and lowest values for each group: **1 mark**

Midpoint × frequency for each group: **1 mark**

Totals for 2nd and 4th columns: **1 mark**

Estimated mean = 24 625 ÷ 200 = 123.125 = 123 min (nearest minute)

Correct figures used: **1 mark**; *answer:* **1 mark**

(b) Number taking 2 hours or less = 3 + 8 + 15 + 42 = 68. ***Find the number of students:* 1 mark**

Probability = $\frac{68}{200} = \frac{17}{50}$ or 0.34. ***Divide by total frequency:* 1 mark**

(c) The exam contained too many questions.

Reason:

either (i) The mean time to finish was longer than the two hours it was supposed to take. ***Compare the mean to the proposed time of 2 hours:* 1 mark**

or (ii) Only about a third of the students finished in time. ***Use the calculation from part (b):* 1 mark**

- It's essential to use the midpoints, as the actual data values are not known. You have to estimate the contribution to the total that each group makes.
- Be careful to divide the right numbers when you're calculating the mean. A common error is to divide by the number of groups. In this case, it's six, which gives an estimated mean of over 4000!
- When you have to make a deduction about the situation in part (c), don't just 'give an opinion', use the facts you've calculated to make a reasoned judgement.

>> Further question

1 The following table gives the speed of cars passing a radar checkpoint.

speed (s km/h)	frequency (f)
$0 \leqslant s < 40$	6
$40 \leqslant s < 60$	8
$60 \leqslant s < 70$	26
$70 \leqslant s < 80$	82
$80 \leqslant s < 100$	103
$100 \leqslant s < 150$	48

(a) Which is the modal speed class?

(b) How many cars' speeds were recorded altogether?

(c) In which class is the median speed?

(d) Calculate an estimate of the mean speed of the cars. Give your answer to the nearest km/h.

>> Answer to further question

1 (a) $80 \leqslant s \leqslant 100$ (b) 273 (c) $80 \leqslant s < 100$

(d)

speed (s km/h)	frequency (f)	mid-interval values (m)	mf
$0 \leqslant s < 40$	6	20	120
$40 \leqslant s < 60$	8	50	400
$60 \leqslant s < 70$	26	65	1690
$70 \leqslant s < 80$	82	75	6150
$80 \leqslant s < 100$	103	90	9270
$100 \leqslant s < 150$	48	125	6000
totals	**273**	–	**23 630**

Estimated mean = 23 630 ÷ 273 = 88.55677... km/h = 89 km/h to nearest km/h.

Complete the facts

Number

Fractions, decimals and percentages

1 Moving the digits in a number one place to the left multiplies the number by …

2 The top and bottom numbers in a fraction are called the …

3 To change a fraction to a decimal, you …

4 A percentage is …

5 To round a number to 3 significant figures, you …

6 To remember the correct order for operations, use the 'word' …

Negative numbers

1 Adding a negative number is the same as …

2 Subtracting a negative number is the same as …

3 When you multiply or divide two negative numbers, the answer is …

Factors and multiples

1 A factor of a number is …

2 To create multiples of a number, you …

3 Prime numbers are …

4 To write a number as a product of its prime factors, draw a …

5 HCF and LCM stand for …

Powers, roots and standard index form

1 The large and small numbers that form a power are called …

2 To multiply two powers of the same base, you …

3 The opposites of squaring and cubing are …

4 The two parts of a number in standard index form are …

5 In standard form numbers less than 1, the index is …

Ratio and proportion

1 A ratio written in the form $1 : n$ is called a …

2 Ratios in their simplest form contain …

3 The steps in carrying out a proportional division are …

4 When two quantities are in direct proportion, they …

5 The unitary method involves …

Algebra

Algebraic expressions

1 Expressions are built up from …

2 In an expression, ab means …

3 Substitution is …

4 A formula is …

Equations and inequalities

1 An equation is …

2 When solving an equation, it is essential to …

3 The difference between $x \leqslant 2$ and $x < 2$ is …

4 When you multiply or divide an inequality by a negative number, remember to …

Manipulating expressions

1 An identity is …

2 The quantity being calculated in a formula is called its …

3 When multiplying a bracket by a term, you need to …

4 The opposite of expanding a bracketed expression is called …

5 When multiplying brackets containing two terms, the number of terms in the expansion is …

Sequences

1 A sequence is …

2 The two types of rule for sequences are …

3 u_n represents …

4 1, 4, 9, 16, 25, 36 are the first six …

5 1, 3, 6, 10, 15, 21 are the first six …

6 1, 8, 27, 64, 125, 216 are the first six …

7 If the term-to-term rule is 'add 5', the position-to-term rule contains …

Co-ordinates and graphs

1 A co-ordinate grid with positive and negative numbers on the axes is divided into four …

2 Horizontal lines have equations of the form …

3 Vertical lines have equations of the form …

4 In the equation $y = mx + c$, m and c stand for …

5 Lines with the same gradient are …

6 The number of points you should plot for a straight-line graph is at least …

7 How many solutions can a quadratic equation have? …

Shape, space and measures

Measures and dimensions

1 In the metric system, milli-, centi- and kilo- mean …

2 An interval is …

3 To calculate average speed, you …

4 The dimension of an expression tells you …

5 A measurement given to the nearest metre may be out by up to …

Shapes and angles

1 Types of angles that add up to 180° include …

2 Angles that add up to 360° are …

3 Types of angles that are equal include …

4 The interior angle sum of a polygon with n sides is …

5 Polygons with all sides and angles equal are called …

6 A net is …

7 The three 'views' of a solid shape are …

8 A prism has a uniform …

9 A 'point' of a pyramid is called its …

Pythagoras' Rule

1 The longest side in a right-angled triangle is called the …

2 Pythagoras' Rule states that …

3 In a rectangle, you can use Pythagoras' Rule to find the length of the …

4 On a coordinate grid, you can use Pythagoras' Rule to find the …

Mensuration

1 The area of a rectangle is given by …

2 The area of a triangle is given by …

3 π is approximately …

4 To calculate the circumference of a circle, you …

5 The area of a circle of radius r is …

6 To find the volume of a cuboid, you …

7 To find the volume of a prism, you …

Constructions and loci

1 A locus is …

2 The locus of points that are a fixed distance from a fixed point is …

3 The locus of points that are a fixed distance from a fixed line is …

4 The locus of points that are equidistant from two fixed points is …

5 The locus of points that are equidistant from two fixed lines is …

6 To construct a triangle with sides of 5 cm, 6 cm and 7 cm, you …

Transformations

1 The shapes before and after a transformation are called …

2 A vector is …

3 In a reflection, corresponding points on the object and image are …

4 In a rotation, the point that doesn't move is called …

5 The size of an enlargement is determined by …

Handling data

Statistical diagrams

1 In a pie chart, the angle for one item is …

2 In a stem-and-leaf diagram, the stems and leaves are …

3 In a graph of distance against time, the gradient of the line tells you …

Statistical calculations

1 The mean of a set of data is …

2 To find the median of a set of data, you have to …

3 The mode means …

4 The difference between the largest and smallest values is called …

5 When data are more consistent, the range is …

Probability

1 An event can have several …

2 The range of values allowed for probability is between …

3 Relative frequency is …

4 To analyse probabilities for two events, the two types of diagram you can use are …

5 To find the probability for one event and another, you …

6 To find the probability for one event or another, you …

Scatter diagrams and correlation

1 If points on a scatter diagram are close to the line of best fit, the correlation is …

2 If as one value increases, the other decreases, the correlation is …

3 If data are positively correlated, the empty areas of the diagram are …

4 Do not use a scatter diagram to make predictions …

Complete the facts – answers

Number

Fractions, decimals and percentages

1 10

2 numerator and denominator

3 divide numerator by denominator

4 a fraction with denominator 100

5 look at the third digit to see which place it is in (hundreds, units, tenths, etc.), then round to this place

6 BIDMAS

Negative numbers

1 subtracting a positive number

2 adding a positive number

3 positive

Factors and multiples

1 anything that divides into the number with no remainder

2 multiply the number by 1, 2, 3, …

3 only divisible by themselves and 1

4 factor tree

5 highest common factor and lowest common multiple

Powers, roots and standard index form

1 base and index

2 add the indices

3 square root and cube root

4 a number between 1 and 10 (the significant digits) and a power of 10

5 negative

Ratio and proportion

1 unitary ratio

2 the smallest possible whole numbers

3 add the numbers in the ratio to find out how many parts are needed altogether, divide the amount by this, then multiply by each number in the ratio

4 are in a fixed ratio

5 reducing quantities to 1, then multiplying

Algebra

Algebraic expressions

1 terms

2 a times b

3 replacing letters by their values

4 a set of instructions for calculating something

Equations and inequalities

1 a puzzle with a definite solution

2 do the same thing to both sides

3 $x = 2$ is allowed in the first, but not in the second

4 reverse the direction of the inequality

Manipulating expressions

1 true for all values you substitute into it

2 subject

3 multiply all terms in the bracket by the term outside

4 factorising

5 4

Sequences

1 a set of numbers that follows a rule

2 term-to-term and position-to-term

3 the nth term

4 square numbers

5 triangular numbers

6 cube numbers

7 $\times 5$

Co-ordinates and graphs

1 quadrants

2 y = a number

3 x = a number

4 gradient and y-intercept

5 parallel

6 3

7 2, 1 or 0

Shape, space and measures

Measures and dimensions

1 $\times \frac{1}{1000}$, $\times \frac{1}{100}$ and $\times$ 1000

2 the difference between two times

3 divide distance by time taken

4 whether it represents area, volume, etc.

5 half a metre

Shapes and angles

1 interior angles of a triangle, adjacent angles, allied angles

2 interior angles of a quadrilateral, angles round a point, exterior angles of any polygon

3 vertically opposite angles, corresponding angles, alternate angles

4 $(180n - 360)°$

5 regular

6 a 'map' of all the faces of a solid shape

7 plan, front elevation, side elevation

8 cross-section

9 apex

Pythagoras' Rule

1 hypotenuse

2 $h^2 = a^2 + b^2$

3 diagonal

4 distance between two points

Mensuration

1 length $\times$ width

2 $\frac{1}{2}$ base $\times$ height

3 3.142

4 multiply diameter by π

5 πr^2

6 multiply length $\times$ width $\times$ height

7 multiply area of cross-section $\times$ length

Constructions and loci

1 a set of points that follow a rule

2 a circle

3 two parallel lines

4 the perpendicular bisector of the points

5 the angle bisector of the lines

6 draw a base line, then use your compass to locate the third vertex

Transformations

1 object and image

2 a way of describing a translation

3 equidistant from the mirror line

4 the centre of rotation

5 the scale factor

Handling data

Statistical diagrams

1 360° ÷ total frequency

2 usually, tens and units

3 speed

Statistical calculations

1 the total divided by the number of data items

2 put the values in order

3 the most frequent

4 the range

5 smaller

Probability

1 outcomes

2 0 and 1

3 frequency of outcome ÷ number of trials

4 possibility space tables and tree diagrams

5 multiply the probabilities

6 add the probabilities

Scatter diagrams and correlation

1 strong

2 negative

3 top left/bottom right

4 outside the range of the data

Answers to practice questions

Number

The decimal number system (p. 3)

1 (a) $25 \times 3 = 75 \rightarrow 7.5$

(b) $3 \times 12 = 36 \rightarrow 0.36$

(c) $64 \div 8 = 8 \rightarrow 0.8$

(d) $144 \div 3 = 48 \rightarrow 48$

2 (a) (i) 61 (ii) 61.3 (iii) 61.25

(b) (i) 600 (ii) 590 (iii) 588.6

3 (a) $(30 - 20) \times 2 = 20$: 35.4

(b) $\frac{10 \times 40}{10} = 40$: 34.3

(c) $6^2 = 36$: 34.7

(d) $\frac{4}{2} \times \frac{2}{4} = 1$: 1.11

Order of operations (p. 5)

1 (a) 25 **(b)** 8 **(c)** 15

(d) 20 **(e)** 130 **(f)** 11

2 (a) 32 **(b)** 5 **(c)** 16

(d) 57 **(e)** 81 **(f)** 64

3 (a) $(4 + 2) \times 3 = 18$

(b) $12 \div (6 - 3) = 4$

(c) none needed

(d) $(5 + 6) \times (6 + 5) = 121$

(e) none needed

(f) $(1 + 2) \times 3 \div (4 + 5) = 1$

4 (a) 350.35 **(b)** 1.48 (3 sf)

(c) 3.46 (3 sf) **(d)** 225

(e) 37 349.4276 **(f)** −1000

Negative numbers (p. 7)

1 (a) 9°C **(b)** 9°C

2 (a) −7 **(b)** 2

(c) 4 **(d)** −14

(e) −32 **(f)** 21

(g) 8 **(h)** 100

3 (a) −134 **(b)** −58

(c) −649 **(d)** −2.3

(e) −7650 **(f)** −50

(g) 0.04 **(h)** 0.0025

Factors and multiples (p. 9)

1 (a) 1, 2, 3, 4, 6, 9, 12, 18, 36

(b) 1, 3, 9, 27

(c) 1, 2, 5, 10, 25, 50

(d) 1, 2, 3, 5, 6, 10, 15, 30

2 (a) 1, 2, 3, 6 **(b)** 1, 3

(c) 1, 2, 5, 10 **(d)** 1

3 (a) 6, 12, 18, 24, 30, 36, 42, 48, 54, 60

(b) 10, 20, 30, 40, 50, 60, 70, 80, 90, 100

(c) 25, 50, 75, 100, 125, 150, 175, 200, 225, 250

(d) 35, 70, 105, 140, 175, 210, 245, 280, 315, 350

4 (a) 30, 60, … **(b)** 175, 350, …

(c) 70, 140, … **(d)** 150, 300, …

5 (a) $2 \times 2 \times 3 \times 3$ (or $2^2 \times 3^2$)

(b) $2 \times 3 \times 7$

(c) $2 \times 2 \times 2 \times 5$ (or $2^3 \times 5$)

(d) $3 \times 3 \times 7$ (or $3^2 \times 7$)

6 (a) HCF: $2 \times 3 = 6$
LCM: $2^2 \times 3^2 \times 7 = 252$

(b) HCF: $2 \times 2 = 4$
LCM: $2^3 \times 3^2 \times 5 = 360$

(c) HCF: 7
LCM: $2 \times 3^2 \times 7 = 126$

(d) HCF: 1
LCM: $2^3 \times 3^2 \times 5 \times 7 = 2520$

Working with fractions (p.11)

1 (a) $\frac{3}{6}$ **(b)** $\frac{9}{12}$ **(c)** $\frac{5}{8}$ **(d)** $\frac{4}{9}$

2 (a) $3\frac{3}{4}$ **(b)** $3\frac{1}{7}$ **(c)** $10\frac{1}{5}$ **(d)** $9\frac{1}{9}$

3 (a) $\frac{13}{3}$ **(b)** $\frac{97}{8}$ **(c)** $\frac{88}{9}$ **(d)** $\frac{111}{4}$

4 (a) $\frac{11}{12}$ **(b)** $\frac{9}{10}$ **(c)** $\frac{2}{3}$ **(d)** $1\frac{1}{2}$

Fractions, decimals and percentages (p. 13)

1

	fraction	decimal	percentage
(a)	$\frac{3}{5}$	0.6	*(60%)*
(b)	$\left(\frac{9}{10}\right)$	0.9	90%
(c)	$\frac{7}{20}$	*(0.35)*	35%
(d)	$\left(\frac{1}{20}\right)$	0.05	5%
(e)	$\frac{11}{25}$	*(0.44)*	44%
(f)	$\frac{1}{8}$	0.125	*(12.5%)*
(g)	$\left(\frac{3}{16}\right)$	0.1875	18.75% $(18\frac{3}{4}\%)$
(h)	$\frac{1}{500}$	0.002	*(0.2%)*
(i)	$2\frac{3}{8}$	*(2.375)*	237.5%
(j)	$\frac{9}{125}$	*(0.072)*	7.2%
(k)	$\left(\frac{1}{9}\right)$	$0.\dot{1}$	$11.\dot{1}\%$ $(11\frac{1}{9}\%)$
(l)	$1\frac{9}{20}$	1.45	*(145%)*

Powers and roots (p. 15)

1 (a) 512 **(b)** 484

(c) 15 625 **(d)** $\frac{1}{32}$ or 0.031 25

(e) 8 **(f)** 2.65 **(g)** 3 **(h)** 2.15

2 (a) 5 **(b)** 10 **(c)** 5 **(d)** −3
(e) 8 **(f)** 9 **(g)** 2 **(h)** 10

Standard index form (p. 17)

1 (a) 2×10^4 **(b)** 4×10^6

(c) 5.5×10^5 **(d)** 9.71×10^{10}

(e) 5×10^{-2} **(f)** 3×10^{-7}

(g) 7.2×10^{-4} **(h)** 3.55×10^{-1}

2 (a) 60 000 000 **(b)** 323 000

(c) 19 000 000 000 **(d)** 40

(e) 0.000 07 **(f)** 0.000 000 0199

(g) 0.0903 **(h)** 0.000 000 0008

3 (a) 8×10^7 **(b)** 5×10^1

(c) 4.9×10^4 **(d)** 4.89×10^8

4 (a) 2.25×10^{10} **(b)** 7.01×10^{10}

(c) -1.004×10^3

(d) 2.5×10^{-4}

Ratio and proportion (p. 19)

1 (a) 5 : 1 **(b)** 3 : 2 **(c)** 3 : 4 **(d)** 5 : 6

2 (a) 1 : 2.5, 0.4 : 1 **(b)** 1 : 1.25, 0.8 : 1

3 (a) £60 : £90 **(b)** £135 : £15

4 125 calories **5** 0.48 m = 48 cm

Percentage calculations (p. 21)

1

	£40	**£22.50**	**15 cm**	**3000 people**
25% of ...	£10	£5.63	3.75 cm	750
7% of ...	£2.80	£1.58	1.05 cm	210

2

	£200	**£11**	**25 litres**	**500 tonnes**
Increase by 3% ...	£206	£11.33	25.75 l	515 t
Decrease by 20% ...	£160	£8.80	20 l	400 t

3 (a) 70% **(b)** 15%

4 (a) 20% increase **(b)** 26% decrease

Algebra

Algebraic expressions (p. 23)

1 (a) 2 **(b)** 1 **(c)** 3 **(d)** 1 **(e)** 1 **(f)** 1
(g) 3 **(h)** 1 **(i)** 1 **(j)** 4

2 (a) equation **(b)** identity **(c)** equation
(d) formula **(e)** equation **(f)** formula
(g) identity **(h)** equation **(i)** formula **(j)** identity

Formulae and substitution (p. 25)

1 (a) 15 **(b)** 42 **(c)** 2 **(d)** 0.01 **(e)** 3
(f) 100 **(g)** −32 **(h)** 1600 **(i)** 40 **(j)** 6 (or −6)

2 (a) 0.6 **(b)** 0.2 **(c)** 0.1875 **(d)** 3 **(e)** 0.3

3 (a) $B = \frac{w}{h^2}$ **(b)** $S = 2(lw + wh + hl)$

Solving equations (p. 27)

1 $x = 7$ **2** $y = 9$ **3** $d = 2$

4 $f = 12$ **5** $c = 5$ **6** $h = \frac{1}{2}$

7 $x = 3$ **8** $x = -1$ **9** $k = 6$ **10** $p = \frac{2}{3}$

Trial and improvement (p. 29)

1 2.2 **2** 2.1 **3** 3.5, −0.1 **4** 1.8, −0.8
5 1.7, 0.3, −1.4 **6** 2.3

Rearranging formulae (p. 31)

1 $x = r - 9$ **2** $x = a + z$

3 $x = \frac{16 - 4y}{5}$ **4** $x = \sqrt{\frac{c}{m}}$

5 $x = 4m$ **6** $x = p^2 + pc$

7 $x = \frac{cb}{m}$ **8** $x = \frac{5 - p}{3 + f}$

9 $y = \frac{9 - 2x}{5}$ **10** $y = \frac{x}{2} - 5$

Using brackets in algebra (p. 33)

1 $x = 4$ **2** $x = 10$ **3** $x = -8$

4 $x = 4.5$ **5** $x = -5\frac{1}{3}$ **6** $x = 4$

7 $x = 15\frac{1}{2}$ **8 (a)** $7x(2x + 1)$ **(b)** $9y(4y - 1)$

9 (a) $5y^2(3y^2 + 5)$ **(b)** $20a(5a + b^3)$

Multiplying bracketed expressions (p. 35)

1 $x^2 + 4x + 3$ **2** $x^2 + 9x + 14$

3 $x^2 + 4x + 4$ **4** $x^2 + nx + mx + mn$

5 $x^2 + 6x + 8$ **6** $x^2 - 2ax + a^2$

7 $x^2 - 6x + 9$ **8** $a^2 - 2ax + x^2$

9 $x^2 - 2xy + y^2$ **10** $x^2 - 4$

11 $x^2 - 9y^2$ **12** $36 - x^2$

Inequalities (p. 37)

1 $x \leqslant 5$ **2** $x > -3\frac{1}{2}$ **3** $a \geqslant \frac{1}{6}$

4 $t \geqslant -1\frac{1}{3}$ **5** $d > -5$ **6** $w \leqslant -2\frac{1}{2}$

7 $f \geqslant -8$ **8** $\frac{3}{2} > p > \frac{1}{2}$ **9** $-\frac{1}{5} \leqslant r \leqslant 2$

10 $-15 < t < 9$

Number patterns and sequences (p. 39)

1 (a) 13 **(b)** −2

(c) 15 **(d)** 80

(e) 5, 10 **(f)** 37

2 (a) 4, 14, 24, 34, 44, …

(b) 5, 1, −3, −7, −11, …

(c) 50, 25, 12.5, 6.25, 3.125, …

(d) 3, 5, 11, 29, 83, …

3 (a) 2, 6, 10, 14, …

(b) 11, 14, 19, 26, …

(c) 3, 9, 27, 81, …

(d) 2, 8, 20, 40, …

Sequences and formulae (p. 41)

1 **(a)** First term 6, add 5

(b) $u_n = 5n + 1$

(c) 51

2 **(a)** First term 3, add 6

(b) $u_n = 6n - 3$

(c) 57

3 **(a)** First term 10, subtract 1

(b) $u_n = -n + 11$ or $11 - n$

(c) 1

4 **(a)** First term 1, add 0.2

(b) $u_n = 0.8 + 0.2n$

(c) 2.8

5 **(a)** First term 2, add 3, 4, 5, etc.

(b) $u_n = \frac{n(n+3)}{2}$

(c) 65

6 **(a)** First term 12, multiply by 4

(b) $u_n = 3 \times 4n$

(c) 3 145 728

7 **(a)** First term 0, add 3, 5, 7, etc.

(b) $u_n = n^2 - 1$

(c) 99

8 **(a)** —

(b) $u_n = 2n^3$

(c) 2000

9 **(a)** First two terms 11 and 17, add previous two terms

(b) —

(c) 809

10 **(a)** —

(b) $u_n = \frac{2n-1}{2^n}$

(c) $\frac{19}{1024}$

Co-ordinates (p. 43)

1 A: (1, 1); B: (4, 0); C: (−5, 1); D: (−1, −2); E: (−4, −4); F: (4, −2); G: (−3, 4); H: (0, −3); I: $(3\frac{1}{2}, 3\frac{1}{2})$; J: $(2\frac{1}{2}, -5)$.

2 (1, −3) or (5, −3); any point (3, y) with $y < 0$.

3 There are four types of rectangle with whole-number co-ordinates:

base	height	possible co-ordinates
6	1	(8, 2), (8, 0), (−4, 2), (−4, 0)
3	2	(5, 3), (5, −1), (−1, 3), (−1, −1)
2	3	(4, 4), (4, −2), (0, 4), (0, −2)
1	6	(2, 4), (2, −5), (1, 7), (1, −5)

Others are possible if you use fractions for the base or height!

Lines and equations (p. 45)

1 gradient = 7, y-intercept = 3

2 gradient = 3, y-intercept = −5

3 gradient = $\frac{1}{3}$, y-intercept = 12

4 gradient = 3, y-intercept = −2

5 gradient = −3, y-intercept = 4

6 gradient = 2, y-intercept = −7

7 $y = 4x + 5$ **8** $y = x + 6$ **9** $y = 2x + 1$

Quadratic graphs (p. 47)

1 (a)

x	−5	−4	−3	−2	−1	0	1	2	3	4	5
y	30	21	14	9	6	5	6	9	14	21	30

(b)

x	−5	−4	−3	−2	−1	0	1	2	3	4	5
y	−38	−26	−16	−8	−2	2	4	4	2	−2	−8

(c)

x	−5	−4	−3	−2	−1	0	1	2	3	4	5
y	7.5	4	1.5	0	−0.5	0	1.5	4	7.5	12	17.5

(d)

x	−5	−4	−3	−2	−1	0	1	2	3	4	5
y	63	42	25	12	3	−2	−3	0	7	18	33

2 (a)

x	−3	−2	−1	0	1	2	3	4	5
y	14	7	2	−1	−2	−1	2	7	14

Solutions: $x = 2.4$ and $x = -0.4$

(b)

x	−3	−2	−1	0	1	2	3
y	17	2	−7	−10	−7	2	17

Solutions: $x = 1.8$ and $x = -1.8$

Shape, space and measures

Measures and accuracy (p. 49)

1 (a) 3000 m **(b)** 5 cl **(c)** 4200 g **(d)** 0.312 l **(e)** 7000 kg
(f) 200 mm² **(g)** 1.14 l **(h)** 57.24 kg **(i)** 10.6 oz (3 sf)
(j) 125 mm **(k)** 150 mi **(l)** 22.0 lb (3 sf)

2 (a) 9.5 cm ⩽ length < 10.5 cm; 4.5 cm ⩽ length < 5.5 cm;
42.75 cm² ⩽ area < 57.75 cm²

(b) 295 m ⩽ length < 305 m; 245 m ⩽ length < 255 m;
72 275 m² ⩽ area < 77 775 m²

(c) 15.75 mm ⩽ length < 16.25 mm;
6.25 mm ⩽ length < 6.75 mm;
98.437 5 mm² ⩽ area < 109.687 5 mm²

Dimensions (p. 51)

1 (a) $2\pi r$ length **(b)** πr^2 area **(c)** $\pi r^2 h$ volume **(d)** $2\pi rh$ area

2 There are too many dimensions. For it to be area, it must have only two dimensions.

3 It must represent a volume if it is of dimension three.

Time (p. 53)

1

12-hour	*(3pm)*	*(2.25pm)*	6.40 am	10.10 pm	*(4.45 am)*	12.05 am	*(7.15pm)*
24-hour	*(15:00)*	14:25	*(06:40)*	*(22:10)*	04:45	*(00:05)*	19:15

2 (a) 180 min **(b)** 720 sec **(c)** 150 min = 9000 sec

(d) 255 min = 15 300 sec **(e)** 8 min 20 sec **(f)** 7 h 28 min

(g) 126 min **(h)** 948 sec

3

start	13:05	09:40	10.50 pm	16:22
end	16:25	11:15	2.05 am	18:18
interval	3 h 20 min	1 h 35 min	3 h 15 min	1 h 56 min

4 (a) 6.15 pm **(b)** 320 sec = 5 min 20 sec

(c) 390 min ÷ 5 = 78 min = 1 h 18 min

Compound measures (p. 55)

1

	distance	time	average speed
(a)	*(30 km)*	*(2 hours)*	15 km/h
(b)	*(4 km)*	*(30 minutes)*	8 km/h
(c)	*(100 km)*	2½ hours	*(40 km/h)*
(d)	12.5 metres	*(0.05 seconds)*	*(250 m/s)*
(e)	28 800 km	*(1 hour)*	*(8 km/second)*
(f)	*(400 m)*	*(1 minute)*	24 km/h

2 Platinum: 1075 g; gold: 1158 g. The gold is heavier by 83 g.

3 Number of seconds in 1 year is roughly 365 × 24 × 3600 = 31 536 000. Rate of consumption is roughly 2.5 tonnes per second.

Angle facts (p. 57)

1 (a) *ab, cd, ef, gh* **(b)** *ac, bd, eg, fh* **(c)** *ce, df*
(d) *ae, bf, cg, dh* **(e)** *cf, de*

2 (a) $a = 90°, b = 90°, c = 117°$

(b) $d = 41°, e = 49°$

(c) $f = 120°$

(d) $g = 75°, h = 75°, i = 65°, j = 40°$

(e) $k = 243°$

Properties of shapes (p. 59)

1 (a) 32° **(b)** 96° **(c)** 71° **(d)** 200°

2 (a) 1800° **(b)** 3240°

3 (a) 140° **(b)** 40° **(c)** 9

Pythagoras' Rule (p. 61)

1 (a) 10.63 cm (2 dp) **(b)** 23.82 cm (2 dp)

(c) 13.23 cm (2 dp) **(d)** 10.01 cm (2 dp)

2 4.58 m (to 2 dp) **3** 2.97 m (2 dp)

4 (a) 9.9 **(b)** 8.1 **5** Prakash ($15^2 = 12^2 + 9^2$)

Calculating areas (p. 63)

1 17.5 m² **2** 206 m² **3** 3 rolls

4 7 tins **5** 22.75 cm² **6** 10 cm

Circle calculations (p. 65)

1 (a) 43.98 cm **(b)** 12.57 cm

(c) 78.54 cm **(d)** 68.49 cm

2 (a) 50.27 cm² **(b)** 452.34 cm²

3 (a) 15.90 m² **(b)** 21.49 m²

4 5.64 cm

5 (a) Perimeter: $C1 = 4\pi$ cm, $C2 = 12\pi$ cm.
Area: $C1 = 4\pi$ cm², $C2 = 36\pi$ cm².

(b) 3 times **(c)** 9 times

Solid shapes (p. 67)

1 (a) 6 faces, 12 edges, 8 vertices

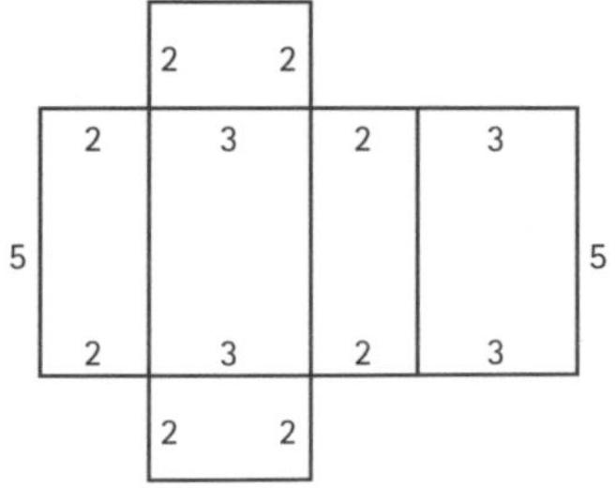

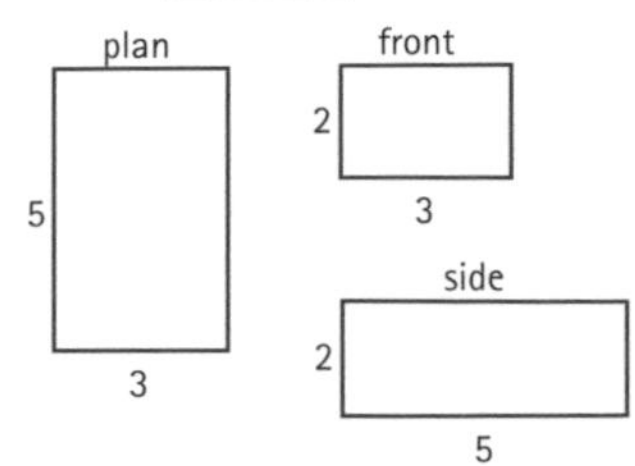

(b) 5 faces, 8 edges, 5 vertices

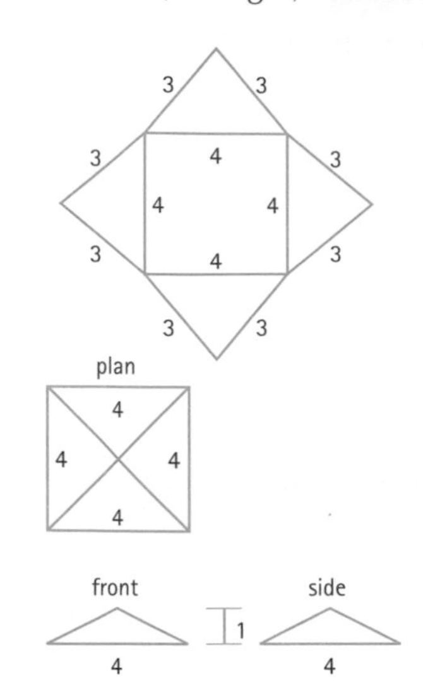

(c) 8 faces, 18 edges, 12 vertices

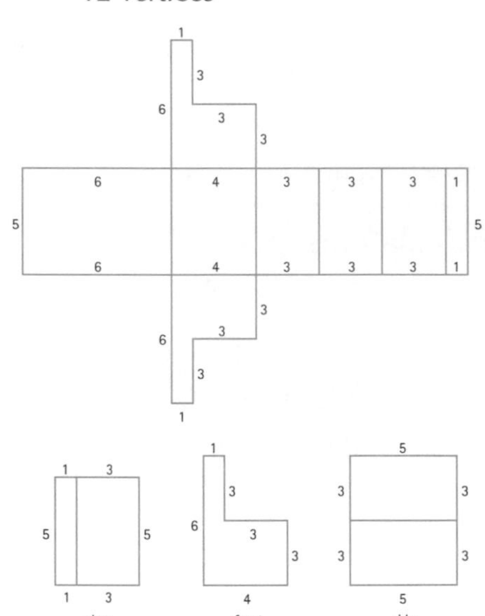

Volume calculations (p. 69)

1 $60\,m^3$ **2** $108\,m^3$

3 $3770\,cm^3$ **4** $0.62\,m^3$

5 $78\,m^3$

Constructions (p. 71)

All answers in this exercise should be self-checking.

Loci (p. 73)

1 (a-b)

Diagram shown is reduced to fit – use for reference only.

RS = 9 cm, RT = 6 cm, ST = 7.5 cm
Circle radii: R = 5 cm, S = 6 cm, T = 3.5 cm

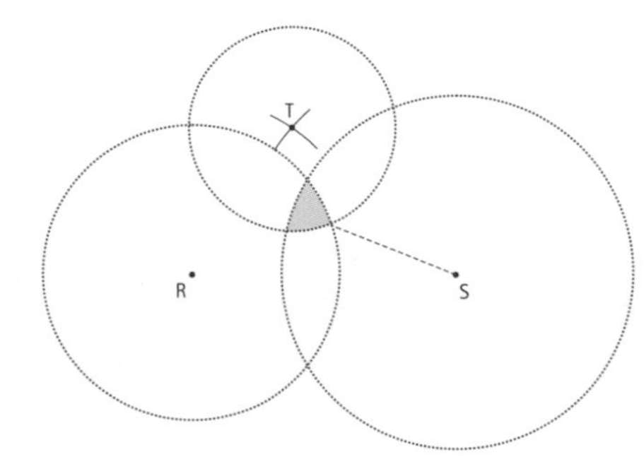

(c) 46 km (allow 2 km either way)

2 (a-b)

Diagram shown is reduced to fit – use for reference only.

Triangle sides 7 cm
Locus distance from triangle = 1.5 cm

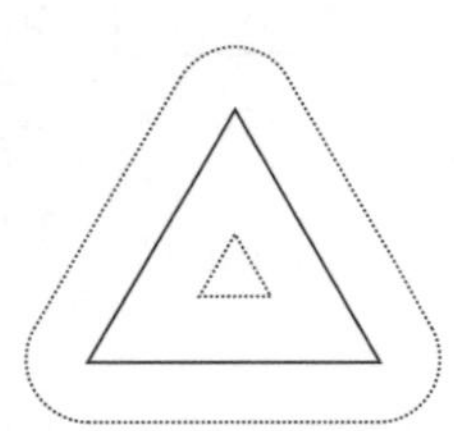

3 (a) $y = 5$ [passes through (0, 5) and (10, 5)]

(b) $y = 10 - x$ [passes through (0, 10) and (10, 0)]

(c) (5, 5)

Transformations (p. 75)

1 (a) (i) (4, 6), (4, 8), (3, 8)
(ii) (9, 3), (9, 5), (8, 5)

(b) (i) (6, 0), (6, 2), (7, 2)
(ii) (2, 3), (2, 4), (0, 4)

(c) (i) $(-7, -3), (-5, -3), (-5, -2)$
(ii) $(0, -4), (0, -6), (1, -6)$

(d) (i) $(2, -5), (2, -1), (0, -1)$
(ii) (8, 8), (9, 8), (8, 10)

2 (a) Translation with vector $\begin{pmatrix} -3 \\ 0 \end{pmatrix}$

(b) Reflection in $y = -3$

(c) Rotation 90° anticlockwise, centre $(-2, 1)$

(d) Enlargement, scale factor 4, centre (7, 0)

Handling data

Statistical tables and diagrams (p. 77)

1 (a)

Stem	Leaf
2	9 9
3	
4	5 5 7 7 8 8
5	0 0 1 1 2 3 5
6	0 5
7	0 2 2

Key: 2 | 0 = 20

(b) Bar chart drawn with categories 20–29, 30–39, … 70–79.
Frequencies 2, 0, 6, 7, 2, 3.

2 (a) £9800

(b) 2003; £1500

(c) 42.9%

3 (a) 30 mins

(b) 10 km/h

(c) 6 minutes

(d) $1\frac{1}{3}$ km/h

Pie charts (p. 79)

1

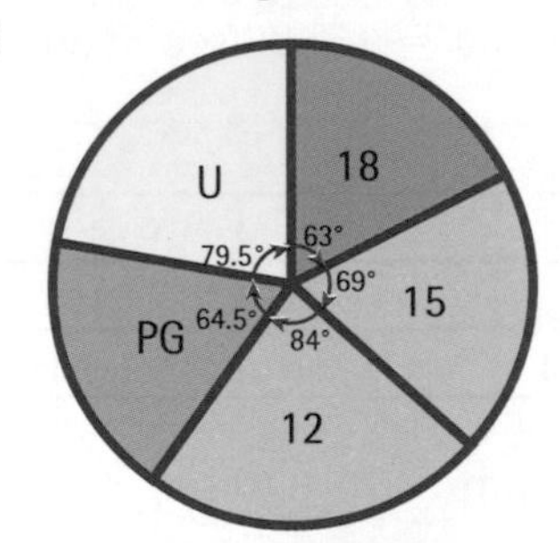

2

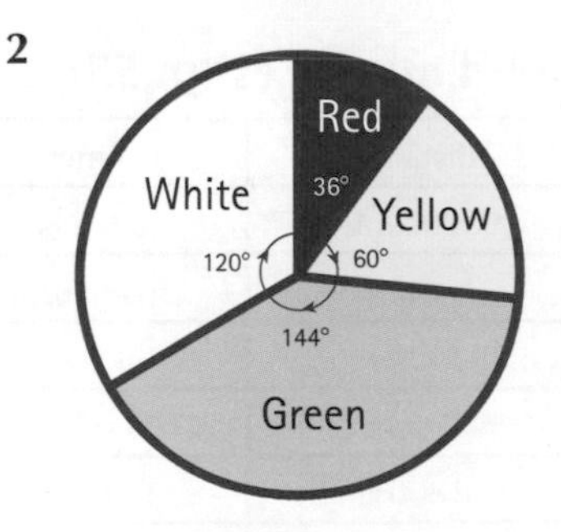

Finding averages (p. 81)

1 Mean value 9 apples (8.95)

2 Modal class is 0–4.

3 1.5

Comparing sets of data (p. 83)

1 (a)

	longest	shortest	mean	range
Markit	1730	1650	1700	80
Skribbla	1930	1640	1750	290

(b) Skribbla: the mean is slightly higher, the shortest is about the same and the longest is quite a bit longer. The poorer consistency turns out to be a good thing!

2 (a) 139 in total

(b) The Gromore results are slightly more consistent and have a higher mode than the Sproutwell plants, so Gromore appears to be better.

Probability (p. 85)

1 Total number of cars counted = 144 + 75 + 36 + 33 + 12 = 300

(a) $P(1) = \frac{144}{300} = \frac{12}{25}$ or 0.48

(b) $P(\geqslant 3) = \frac{36+33+12}{300} = \frac{81}{300} = \frac{27}{100}$ or 0.27

2

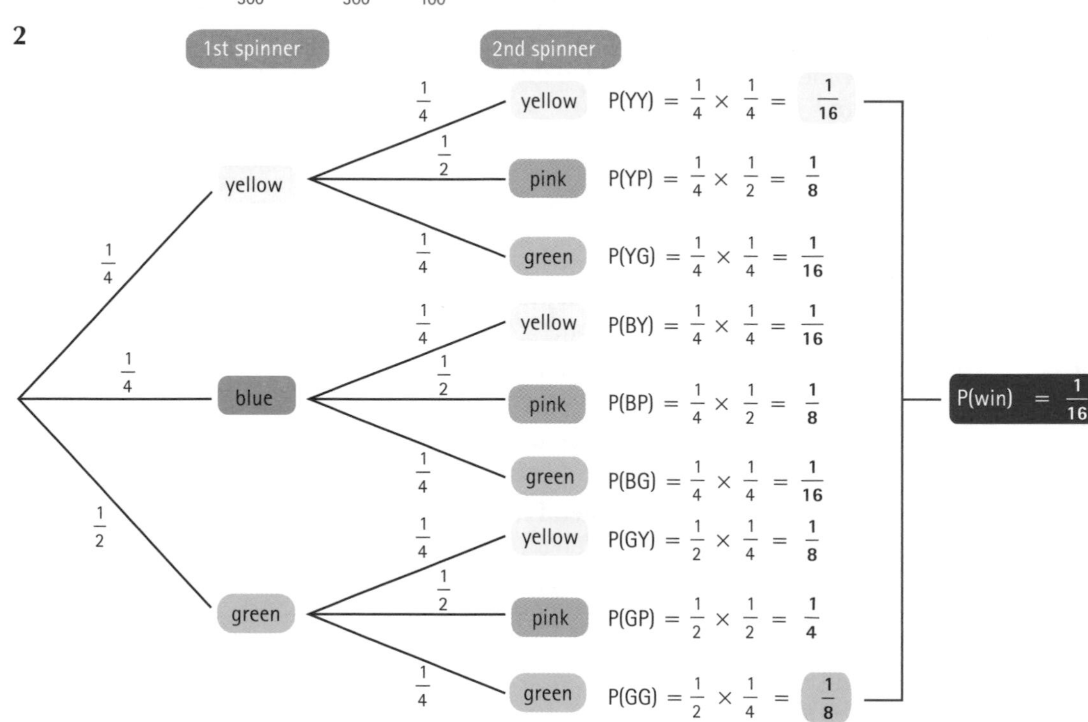

3 (a)

		1st die					
		1	2	3	4	5	6
2nd die	1	0	1	2	3	4	5
	2	1	0	1	2	3	4
	3	2	1	0	1	2	3
	4	3	2	1	0	1	2
	5	4	3	2	1	0	1
	6	5	4	3	2	1	0

(b) 1 is most likely: $P(1) = \frac{10}{36}$

(c) $P(2) = \frac{8}{36} = \frac{2}{9}$.

$\frac{2}{9}$ of 100 = 22.22…,

so roughly 22 times.

Scatter diagrams and correlation (p. 87)

1

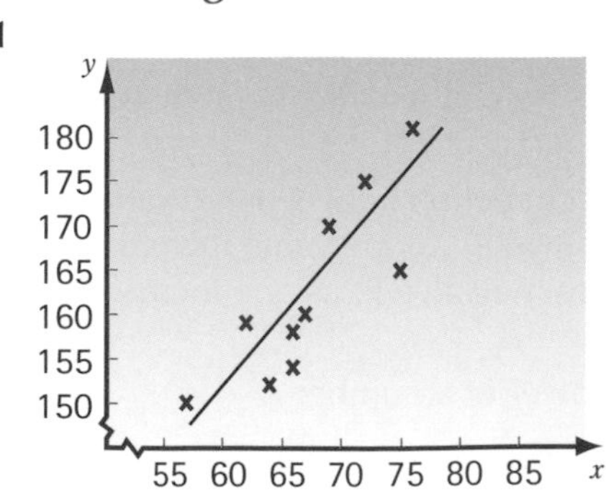

2 There is a strong correlation between height and weight.

3 Nisha's height will be about 152 cm.

Glossary

A

acute angle an angle less than 90° (less than a right angle)

algebra the branch of mathematics that deals with the general case. Algebra involves the use of letters to represent variables and is a very powerful tool in problem-solving.

angle a measure of rotation

apex the 'point' of a pyramid

arc part of the circumference of a circle

area a measure of two-dimensional space

average see mean, median and mode

B

bar chart a frequency chart, where the frequency of the data is proportional to the height of the bar

base the large number in a power – the number being repeatedly multiplied

bearing a measure of direction used in navigation

BIDMAS a made-up 'word' to help remember the order of operations – **B**rackets, **I**ndices, **D**ivision/**M**ultiplication, **A**ddition/**S**ubtraction

C

centre of enlargement/rotation the point that is fixed during enlargements/rotations

chord a line across a circle, not passing through the centre

circumference the complete boundary of a circle

coefficient the number in front of the variable that multiplies it, for example in $2x$ the coefficient of x is 2 and in $5y$ the coefficient of y is 5

common factor a factor of two or more numbers

common multiple a multiple of two or more numbers

composite number any number that is not prime

compound interest the type of interest that is paid by most banks, where the interest is added to the principal invested and then in subsequent years interest is earned by the original interest

compound unit a unit built from more than one basic unit, such as m/s (speed) or kg/m^3 (density)

construction precise mathematical drawing using a compass, ruler and protractor

correlation the connection between two variables. Correlation is positive if, as one value increases, the other increases; negative if, as one increases, the other decreases.

cube the third power of a number

cube number the sequence of cube numbers is 1, 8, 27, 64, 125, etc. It is made up from $1 \times 1 \times 1$, $2 \times 2 \times 2$, $3 \times 3 \times 3$ and so on.

cube root the opposite of cubing

cuboid a rectangular prism, having six rectangular faces

cylinder a prism with a circular cross-section

D

decomposition writing a number as the product of its prime factors, e.g. $120 = 2^3 \times 3 \times 5$

denominator the lower part of a fraction

diameter the distance across a circle from one point on the circumference to another point on the circumference, passing through the centre

dimension the type of quantity represented by an expression. In the exam, you are asked to distinguish between expressions for length, area and volume.

direct proportion when two quantities are in a fixed ratio to one another: when you double one, you double the other, etc.

distribution a set of data values, together with their frequencies

dodecahedron a regular solid with 12 faces

E

elevations views of a solid object from the front and side

equation a mathematical statement, usually in algebra, where two sides of the statement are equal. The aim is to work out the value of the unknown by manipulating the equation.

equation, linear an equation where the highest power of any of the terms is 1

equation, quadratic an equation where the highest power of any of the terms is 2

equilateral triangle a triangle that has all three sides of equal length and all three angles equal 60°

estimate (1) the process of comparing the size of a property of an object with a known quantity

estimate (2) perform a rough calculation using rounded numbers

F

factor a number or variable that divides into other numbers or variables without a remainder

factorisation the process of extracting the highest common factors from an expression

formula an algebraic statement that is the result of previously established work. A formula is accepted to be correct and can be used in subsequent work. For example, $A = \pi r^2$ is accepted as the formula for the area of a circle. It is not necessary to prove this formula every time we need to use it.

function a rule which takes a set of input values and generates a set of output values – a 'number machine'

G

generalising a process in mathematical thinking, where a general rule, usually expressed in algebra, is determined

gradient the measurement of the steepness of a line. Where a line slopes from bottom left to top right, the gradient is positive. Where the line slopes from top left to bottom right, the gradient is negative.

graph a visual representation of a set of data, or a relationship between sets of data

H

highest common factor the highest factor that will divide exactly into two or more numbers

hypotenuse the longest side of a right-angled triangle.

I

identity an algebraic statement where the two sides are equal regardless of the values you substitute; for example $2x = x + x$

image the result of a transformation acting on an object shape

imperial units the traditional units of measure which were once used in the UK and the USA, such as feet, inches, pints and gallons

improper fraction a fraction where the numerator is of a higher value than the denominator

inclusive inequality an inequality in which the endpoint is included, e.g. $x \geqslant 5$

independent event An event whose outcome is not dependent on the outcomes of other events.

index the small number in a power – how many copies of the base to multiply together

inequality an algebraic statement in which the two sides are unequal

integer a positive or negative whole number

intercept the position where a graph crosses the y-axis

irrational number a number that cannot be written as a fraction, for example $\sqrt{2}$

isosceles triangle a triangle with two equal sides and two equal angles

L

line of symmetry a line that bisects a shape so that each part of the shape reflects the other

locus a set of positions governed by a mathematical rule. Loci may be points, lines, curves or areas.

lowest common multiple the lowest number that two or more numbers will divide into; for example the lowest common multiple of 4 and 8 is 8, because 8 is the lowest number that both 4 and 8 will go into

M

mean the arithmetic average, which is calculated by adding all of the items of data and then by dividing the answer by the number of items of data

median the middle value in a set of data which are in order of size

metric the system of measurements usually used in Europe and the most commonly used system in science

mode the value that occurs the most often in a set of data

N

negative number a number with a value less than zero

net the pattern made by a three-dimensional shape when it is cut into its construction template and then laid flat

numerator the upper part of a fraction

O

object 'original' shape, before a transformation is applied

obtuse refers to any angle greater than 90° but less than 180°

outcome one possible result of a chance event

P

parallel lines lines that are equidistant along the whole of their length, in other words they are a constant distance apart along the whole of their length

perimeter the total distance around the boundary of a shape

perpendicular lines lines that meet at right angles

pi (π) the ratio of the circumference of a circle to its diameter, approximately 3.142

pie chart a chart in the shape of a circle, where the size of the sector shows the frequency

plan a view of a solid object from above

polygon a many-sided shape

position-to-term rule a formula that calculates the terms (u_n) of a sequence from their positions (n)

powers a way of writing repeated multiplication. The base number is multiplied repeatedly – the index tells you how many copies of the base number to use, e.g. $5^3 = 5 \times 5 \times 5$.

prefix a short word added to the front of a unit to change its size, e.g. milli-, centi-, kilo-, etc.

prime number a number with two and only two factors. The factors are the number itself and 1.

prism a solid with a uniform cross-section

probability a number between 0 and 1 that represents the chance that an outcome will occur

proportional division dividing an amount in a given ratio

Pythagoras' Rule (theorem) the statement of a relationship between the three sides of a triangle, which is 'the square of the hypotenuse of a right-angled triangle is equal to the sum of the squares of the other two sides'. Using algebra, it is written as $h^2 = a^2 + b^2$.

Q

quadrant one of the four 'quarters' of a co-ordinate grid

quadratic equation an equation where the highest power is 2

quadrilateral a plane four-sided shape

R

radius the distance from the centre of a circle to any point on the remaining part of the circumference. $2 \times$ radius $=$ diameter.

ratio the relationship between one quantity and another

rational number a number that can be written as a fraction

reflex angle any angle greater than 180°.**regular polygon** a plane shape having straight sides of equal length and equal interior angles

right angle any angle that is equal to 90°

rounding reducing the accuracy of a number to a sensible level. Rounding may be to the nearest integer, ten, hundred, etc., to a specified number of decimal places or significant figures.

S

scale factor the number of times a shape is enlarged

scalene triangle a triangle where all of the sides and all of the angles are of different sizes

scatter diagram a plot of two sets of data values as co-ordinates, used to investigate a possible link between them

sequences a set of numbers having a common property. A mathematician can work out what the numbers have in common and so predict further numbers in the sequence.

significant figures or digits the most important digits in a number – those in the places with the highest values, e.g. in 2156 the 2 is the first significant figure as it is in the thousands place

square number the sequence of square numbers is 1, 4, 9, 16, 25, 36, etc. It is made up from 1×1, 2×2, 3×3 and so on.

standard form a way of writing very large or very small numbers, using powers of 10

strict inequality an inequality in which the endpoint is not included, e.g. $x > 5$

subject the quantity calculated by a formula

substitution replacing letters in an algebraic expression by their values

supplementary adding to make 180°

T

tangent a straight line that touches a circle at one point on its circumference

term-to-term rule a formula that calculates the next term of a sequence from the previous term

tolerance the size of the possible error in a measurement, e.g. if $L = 25$ cm, to the nearest cm, then $24.5\text{ cm} \leq L < 25.5\text{ cm}$. The tolerance is 0.5 cm.

translation a sliding transformation

U

unitary ratio a ratio in the form $1 : n$ or $n : 1$

V

VAT Value Added Tax.

vector a quantity describing a translation, e.g. $\begin{pmatrix} 5 \\ -2 \end{pmatrix}$ means '5 units right and 2 down'

volume a measure of three-dimensional space, using cubic units, e.g. cm^3

Last-minute learner

Number

Types of numbers

- **Integers** are positive and negative whole numbers. **Rational numbers** include integers and fractions. Numbers such as $\sqrt{2}$, $\sqrt[3]{10}$ and π are irrational.
- **Multiples** of a number are in its multiplication table. **Common multiples** are multiples of two or more different numbers. Any group of numbers has a **lowest common multiple** (LCM).
- **Factors** of a number are other numbers that divide into it without leaving a remainder. A **prime** number only has two factors: 1 and the number itself, e.g. 2, 3, 5, 7, 11...
  A number that isn't prime is **composite**. Any number can be written as a product of **prime factors**.
- **Common factors** are factors of two or more numbers. Any group of numbers has a **highest common factor** (HCF).

Indices

- **Powers** are made by multiplying a number (the **base**) by itself repeatedly. The **index** tells you how many copies of the base to multiply. Any number to the first power is just the number (e.g. $10^1 = 10$). Any number to the power of zero is 1 (e.g. $5^0 = 1$).
- **Roots** are the opposite or **inverse** of powers. The **square root** of 9 is written $\sqrt{9} = 3$. The **cube root** of 343 is written $\sqrt[3]{343} = 7$.
- **Index laws**: in algebra, $a^n \times a^m = a^{(n+m)}$, $a^n \div a^m = a^{(n-m)}$.
- A negative power is just the **reciprocal** of the positive power. $2^{-2} = \frac{1}{2^2} = \frac{1}{4}$, $2^{-3} = \frac{1}{2^3} = \frac{1}{8}$, etc.
- **Fractional** indices mean roots.
- Numbers in **standard index form** consist of a power of 10 multiplying the significant digits of the number.
  2 million $= 2 \times 1\,000\,000 = 2 \times 10^6$
  $2\,500\,000 = 2.5$ million $= 2.5 \times 10^6$
- Numbers less than 1 need a negative index, e.g. $0.002\,13 = 2.13 \times 10^{-3}$

Fractions

- You can only **add** or **subtract** fractions if they have the same denominator. To add or subtract fractions with different denominators, change to equivalent fractions with equal denominators.
- To **multiply** two fractions together, just multiply the top and bottom numbers separately. Convert mixed fractions to improper ones first. To divide two fractions, invert the second one (turn it upside down) and turn the $\div$ into a $\times$.
- To find $\frac{1}{n}$ of something, divide by n. To find other fractions, first work out $\frac{1}{n}$, then multiply by the numerator.

Proportion and ratio

- There are two main ways to work out a **percentage** of something: either divide the amount by 100 to find 1%, then multiply by the number of per cent; or convert the percentage to a decimal, then multiply the amount by this.
- There are two ways to calculate the result of a percentage **increase or decrease**: work out the amount of change and add/subtract it; or work out the new percentage required and calculate it.
- A **reverse change** is one where you know the amount after a percentage change and want to find the original amount.
- If y is **directly proportional** to x, y and x are in a fixed ratio. Two proportional amounts plotted against each other on a **graph** give a straight line through the origin.
- Amounts that are in direct proportion are also in a constant **ratio**. Ratios can be equivalent to each other, just like fractions.
  So $10 : 4 = 20 : 8 = 100 : 40 = 5 : 2$ in lowest terms.
  Ratios can also be written in unitary form (containing a 1). $10 : 4 = 1 : 0.4 = 2.5 : 1$
- Sometimes you need to divide amounts in a certain ratio. Add up the numbers in the ratio to get the total number of 'shares', find the value of one 'share', then calculate the parts as required.

Algebra

Formulae and expressions

- **Substitution** is replacing letters in a formula, equation or expression by numbers (their **values**). Be careful to evaluate parts of the formula in the correct order.
- **Expressions** in algebra are made up of a number of **terms** added or subtracted together. Each term is made up of letters and numbers multiplied or divided together. Combine **like terms** to simplify an expression.
- A **formula** usually has its **subject** on the left-hand side of the equals sign and an expression on the right-hand side. Any letter in a formula can become the subject by rearranging it. As long as you do the same thing to both sides of your formula, it is still true.

Multiplying and dividing terms

- When **multiplying two terms** together, multiply the numbers first, then multiply the letters in turn, using the index rules.
- When **dividing terms**, write the question in fraction style if it is not already written that way and cancel the numbers as if you were cancelling a fraction to lowest terms, then divide the letters in turn.

Expanding brackets

- When a number or letter multiplies a bracket, **everything** inside the bracket is multiplied. Removing the brackets is called **expansion**.

Factorisation

- **Factorising** is the opposite of expansion. To factorise, look for **common factors** between the terms.
- This process is called **extracting factors**. Sometimes you need to do this in more than one step.
- **Quadratic** expressions can sometimes be factorised into two brackets. First write down a list of the numbers that could be part of the x^2 term. Write down a list of the numbers that could be part of the number term. Test combinations of these numbers to see if you can match the x term in the expression you want to factorise.

Basic equations

- The letter in an **equation** stands for a definite unknown number you have to find.
- To solve an equation:
  - think what you would have to do if you were substituting a number into the equation
  - to find the answer, do the inverse operations in the opposite order.

Trial and improvement

- Sometimes you can't find an exact solution to an equation but can find a reasonable **approximation** using trial and improvement. You use the results of a 'guess' to make better guesses.

Algebraic graphs

- Equations of the form $y = a$, where a is a fixed number, form graphs that are horizontal straight lines. The equation of the x-axis is $y = 0$.
- Equations of the form $x = b$, where b is a fixed number, form graphs that are vertical straight lines. The equation of the y-axis is $x = 0$.
- $y = x$ and $y = -x$ are straight lines through the origin at 45° to the axes.
- All other straight lines (linear graphs) have equations of the form $y = mx + c$. m represents the gradient of the line, a measure of how steep it is. Positive values of m correspond to lines that slope upwards from left to right, negative ones downwards. c represents the y-intercept, the position on the y-axis where the graph crosses it.
- Given any two points, the gradient of the line joining them is given by $\frac{\text{increase in } y}{\text{increase in } x}$.
- Graphs with the same gradient are parallel.
- Quadratic graphs have equations of the form $y = ax^2 + bx + c$. Their 'U' shape is called a parabola.
- To plot a graph, you need a table of x and y values which can be placed onto a grid as co-ordinates. Linear graphs need a minimum of 3 points, curved graphs will usually need more.

Inequalities

- Ranges of numbers are described using inequalities. You can illustrate an inequality on a number line.
- There are four inequality symbols:

$>$ greater than	$\geqslant$ greater than or equal to
$<$ less than	$\leqslant$ less than or equal to

'All the numbers that are 3 or less' is described by the inequality $x \leqslant 3$ (or $3 \geqslant x$).

- Sometimes inequalities can be combined. Suppose that $x < 2$ and $x \geqslant -3$. This makes a **range inequality**, $-3 \leqslant x < 2$.

Sequences

- Sequences are made up of a succession of **terms**. Each term has a **position** in the sequence: 1st, 2nd, etc. A **linear** sequence is one where the difference between terms is always the same number, such as 2, 5, 8, 11, 14. Using n to stand for the position in the sequence and u_n to stand for the term, this sequence has the formula $u_n = 3n - 1$.

Shape, space and measures

Accuracy of measurements

- Suppose the length L of a piece of wire is 67 mm, to the nearest mm. Anything less than 66.5 mm would round down to 66 mm. Anything 67.5 mm and above would round up to 68 mm. So $66.5 \leqslant L < 67.5$.
  If L is given to the nearest 0.1 mm (1 dp) instead, then $66.95 \leqslant L < 67.05$.

Speed, distance and time

- Units of speed are metres per second (m/s), kilometres per hour (km/h), miles per hour (mph), etc.
- Use the '*d-s-t* triangle'. Cover up the letter you want to work out: the triangle gives the formula.

Mensuration – length and distance

- In any right-angled triangle, the **hypotenuse** is the side opposite the right angle.

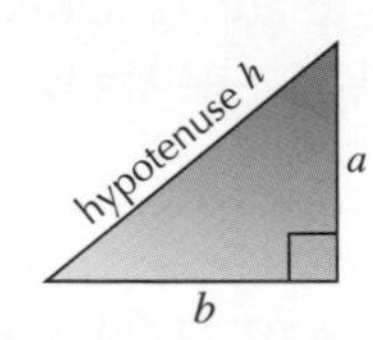

- **Pythagoras' Rule (theorem)**:
  in any right-angled triangle, $h^2 = a^2 + b^2$. Use it to calculate the diagonal of a rectangle, or the distance between two points on a co-ordinate grid.
- The **circumference** of a circle of diameter d (radius r) is $C = \pi d = 2\pi r$. Note that $\pi \approx 3.142$.
- When a wheel turns once (makes one **revolution**), the distance moved by whatever it is attached to (e.g. a

car or a bike) is the same as the circumference of the wheel.

Area

- Using l for length and w for width, the area A of a rectangle is given by $A = lw$.
- Triangles, parallelograms and trapezia all share an important measurement: the **perpendicular height**. These are the area formulae:

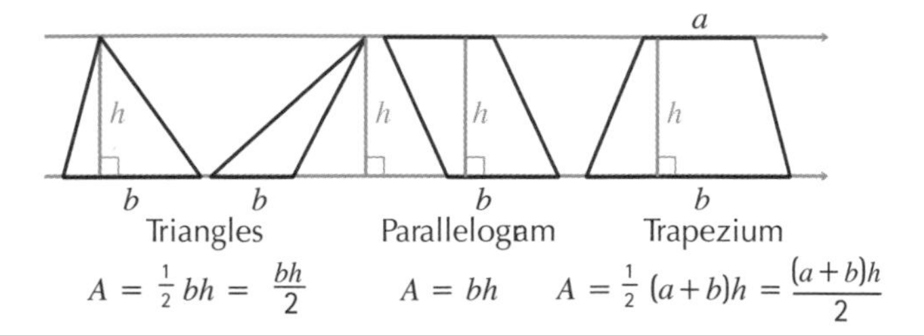

$A = \frac{1}{2}bh = \frac{bh}{2}$ $\quad A = bh \quad$ $A = \frac{1}{2}(a+b)h = \frac{(a+b)h}{2}$

- The area of the circle of radius r is $A = \pi r^2$.

Surface area

- The surface area of a solid object is the combined area of all the faces on the outside. Curved surfaces on spheres, cones and cylinders form part of the surface area too.
- For a prism with cross-section of perimeter P and area A, the total surface area $S = 2A + Pl$. For a cylinder, $S = \pi r^2 + 2\pi rl = 2\pi r(r + l)$.
- Cones need an extra measurement, the slant height s. The curved surface is πrs, so the total surface area is $\pi r^2 + \pi rs = \pi r(r + s)$.
- The surface area of a sphere is $4\pi r^2$.

Volume

- There are four basic volume formulae.

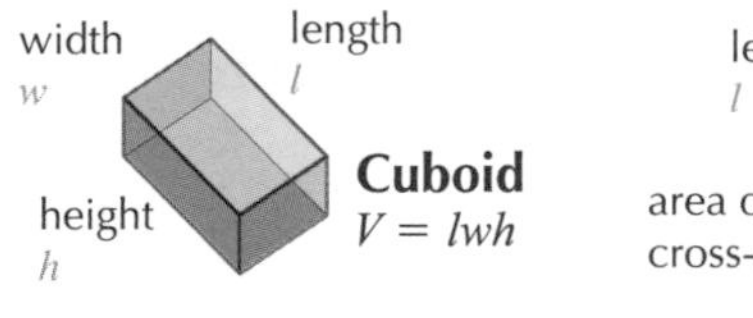

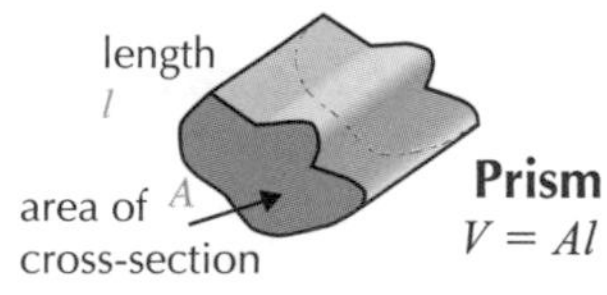

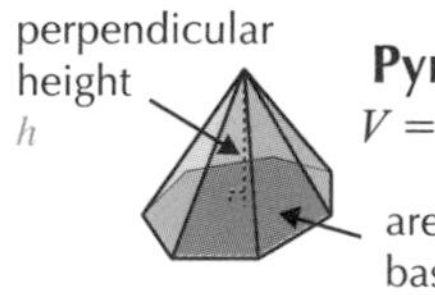

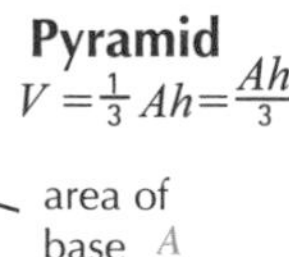

$V = \frac{1}{3}Ah = \frac{Ah}{3}$

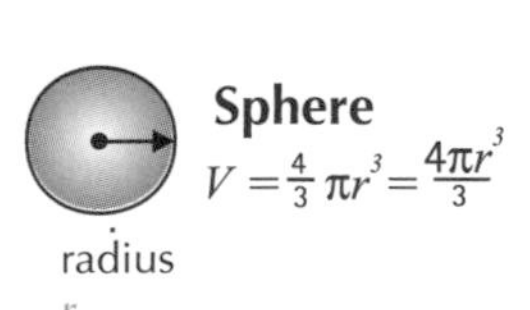

$V = \frac{4}{3}\pi r^3 = \frac{4\pi r^3}{3}$

- For a cylinder, $V = \pi r^2 l$, and for a cone, $V = \frac{1}{3}\pi r^2 h = \frac{\pi r^2 h}{3}$.
- Liquid volume:
  $1\,\text{cm}^3 = 1\,\text{ml}$,
  $1000\,\text{cm}^3 = 1\,\text{l}$,
  $1000\,\text{l} = 1\,\text{m}^3$

Shapes and angles

- Whenever lines meet or **intersect**, the angles they make follow certain rules.

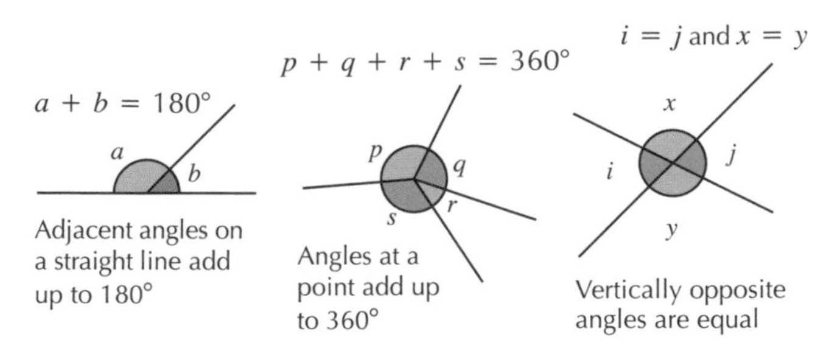

- Three types of relationship between angles are produced when a line called a **transversal** crosses a pair of parallel lines.

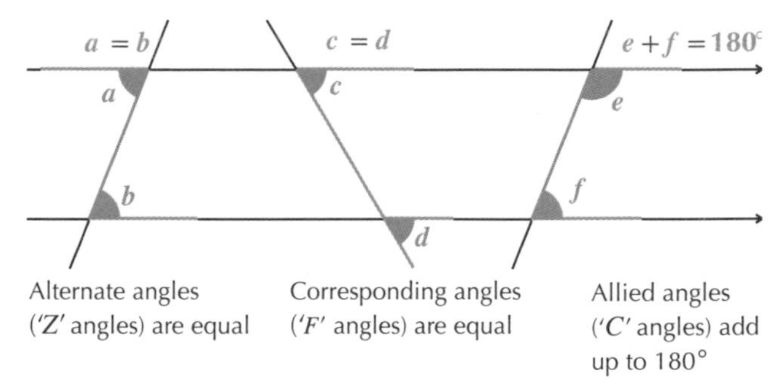

- The **exterior angles** of a **polygon** always add up to exactly 360°.
- Every type of polygon has its own **interior angle sum**. You can calculate it using any of these formulae: n is the number of sides and S is the angle sum.
  $S = (n - 2) \times 180°$ $\qquad$ $S = (180n - 360)°$
  $S = (2n - 4)$ right angles
- Work out the interior angles for **regular** polygons in two ways: work out the angle sum, then divide by the number of sides; or divide 360° by the number of sides to find one exterior angle, then take this away from 180°.

Transformations

- Mathematical transformations start with an original point or shape (the **object**) and transform it (into the **image**).
- A **translation** is a 'sliding' movement, described by a **column vector**, e.g. $\begin{pmatrix} 5 \\ -4 \end{pmatrix}$.
- In a **rotation**, specify an angle and a fixed point called the **centre of rotation**. Given a rotation, you can usually find the centre by trial and error, using a piece of tracing paper.
- When an object is **reflected**, the object and image make a symmetrical pattern. Reflection in any line is possible, but the most likely ones you will be asked to use are these:
  – horizontal lines ($y = a$ for some value of a)
  – vertical lines ($x = b$ for some value of b)
  – lines parallel to $y = x$ ($y = x + a$ for some value of a)
  – lines parallel to $y = -x$ ($y = a - x$ for some value of a).

Enlargement

- To describe an enlargement you need to give a **scale factor** and a **centre of enlargement**. The scale factor affects the **size** of the image. The centre of enlargement affects the **position** of the image.

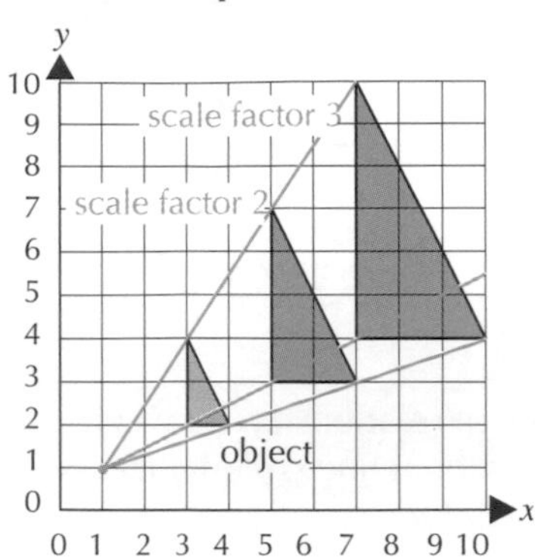

- To find the centre of enlargement, draw lines through corresponding points on the object and image. These all intersect at the centre of enlargement. To find the scale factor, divide the length of a line on the image by the corresponding line on the object.
- The scale factor of a model, map, or scale drawing is usually expressed as a ratio. A 1:40 scale model of a boat is $\frac{1}{40}$ of the actual size of the boat. A diagram of a circuit on a scale of 25:1 is 25 times larger than the real thing.

Congruent and similar shapes

- Two shapes that are identical are **congruent**.
- If the following features of two triangles match, the triangles are congruent: 3 sides (SSS); 2 sides and the included angle (SAS); 2 angles and a side (AAS); in a right-angled triangle, the hypotenuse and one other side (RHS).
- Shapes that are the same apart from their size are **similar**. Similar shapes are **enlargements** of each other. Two triangles are similar if they have identical angles. Their sides are automatically in the same **ratio**.

Loci

- A set of positions generated by a rule is called a **locus**. The four major types are as follows:

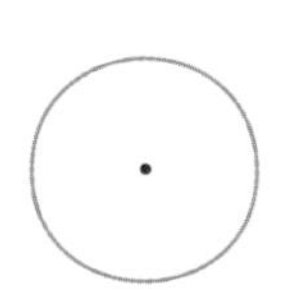

A fixed distance from a fixed point: a circle

A fixed distance from a straight line: two parallel straight lines

Equidistant from two fixed points: the perpendicular bisector of the points

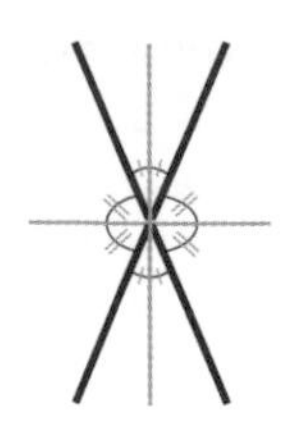

Equidistant from two straight lines: the bisectors of the angles between the lines

- Often you need to combine information from two or more loci to solve a problem. Sometimes this will lead to a region or area, sometimes to a line segment, sometimes to one or more points.

Handling data

Processing data

- The **mode** is the most common value. In the case of data in groups or classes, the group with the highest frequency is called the **modal group** or class.
- The **median** is the middle value in a set, when all the numbers are in order. If you have an even number of data, the median is halfway between the two in the middle.
- The **mean** is the most frequently used average. It is calculated by taking the sum of all the data items, then dividing by the number of items.
- **Range** is the difference between smallest and largest data.

Probability

- **Theoretical probability** is calculated by analysing a situation mathematically. The probability can be used to predict the expected frequency of the outcomes of a number of trials.
- **Experimental probability** is determined by analysing the results of a number of trials of the event.
- The **OR rule**: when two outcomes A and B of the same event are exclusive, $P(A \text{ or } B) = P(A) + P(B)$.
- The **AND rule**: When two events X and Y are independent, $P(X \text{ and } Y) = P(X) \times P(Y)$.

Scatter diagrams and correlation

- When it is thought that two sets of data may be linked, a scatter diagram can be used to investigate. The pairs of data values are plotted as co-ordinates.
- If a link exists, it will be shown in the way the points are distributed on the grid. If the points form a band going roughly from bottom left to top right, there is positive correlation, i.e. as one data item increases, so does the other. If the band goes from top left to bottom right, the data items are negatively correlated.
- A line of best fit can be drawn through the data points. This can be used to predict values by reading across from one scale to the other. It is unwise to do this outside the range of the data, however, as the correlation may not exist there.
- Any 'rogue' data points that lie a long way from the line of best fit are known as outliers and may indicate unreliable data.

Notes